中国苹果品种

Apple Varieties in China

丛佩华　主编

Chief Editor　Cong Peihua

中国农业出版社

China Agriculture Press

序

　　苹果作为世界四大水果之一，地域分布极为广泛。我国作为苹果的发源地之一，具有极为丰富的苹果种质资源。随着苹果产业的不断发展，我国现已拥有渤海湾、黄土高原、黄河故道和西南冷凉高地四大苹果产区，种植品种涉及早、中、晚熟鲜食品种及加工品种。

　　近10年，苹果产业进入快速发展期，2005年统计苹果栽培面积189万hm²，产量2 401.1万t，2013年苹果栽培面积上升到227.2万hm²，产量上升到3 968.3万t。苹果产业已成为涉及一、二、三产业的综合产业，对生态环境、农民增收、企业增效等均有贡献。

　　世界苹果育种的历史已经百年有余，一个世纪以来，育种工作者对苹果新品种的选育和发展贡献良多。纵观全球苹果品种发展格局，国外栽植最为广泛的苹果品种除金冠、元帅、旭等传统品种外，人工选育的富士、嘎拉、乔纳金、粉红女士等品种近年也有较快发展。受苹果消费市场多元化需求的影响，国内外苹果优新品种不断涌现，自20世纪80年代富士苹果引进之后，我国还陆续引进了乔纳金、嘎拉等许多新品种进行栽培。近年来，我国的苹果新品种选育也有一定突破，育成的寒富、秦阳、华硕等多个新品种在各大苹果产区均有不俗的表现。

　　为了让国内外同行对中国现有的栽培苹果品种有所了解，国家苹果产业技术体系7位岗位专家历时3年，筛选现有栽培品种，组织团队成员对选定的144个品种进行统一标准拍摄，并组织编写《中国苹果品种》。该书文字简洁，内容丰富，且每个品种均附有中英文介绍，方便国内外同行查阅，为国内外从事苹果科研及生产相关人员提供了一本很好的参考书。

束怀瑞

2015年6月于山东泰安

PREFACE

Apple (*Malus domestica*) is one of the top four most important fruit trees in the world and is widely cultivated in a variety of geological distribution. China is one of the most important places in which apple was originated and has abundant apple germplasm resources. As the continuous development of apple industry, China has had four main regions of apple production, that is, the Bohai Bay region, the Loess Plateau region in the northwest, the Old Yellow River region, and cold highland region in the Southwest. The varieties cultivated not only include early, mid and late-ripening fresh eating apples but processing apples as well.

Apple industry in China has entered a rapid developmental period in recent 10 years. The apple cultivation area was increased from 1.89 million hectares in 2005 to 2.27 million hectares in 2013, and the total production was increased from 24.011 million tons in 2005 to 39.683 million tons in 2013. Apple industry has become a comprehensive industry including primary, secondary and tertiary industries has contributed a lot to the ecological environment, farmer income and enterprise income.

Apple breeding in the world commenced a century ago. Apple breeders have contributed a lot to the apple industry for breeding and extending new varieties in the world. At present, the most widely cultivated apple varieties in the world include not only the conventional varieties such as Golden Delicious, Delicious and McIntosh, but artificially bred varieties such as Fuji, Gala, Jonagold and Pink Lady as well. The trend of apple breeding is mainly determined by the demands of consumers and markets. Fuji apple was introduced into China in the early 1980s. Many other varieties such as Jonagold and Gala were introduced into China afterwards. Apple breeders in China have made a profound progress in breeding new apple varieties in recent years, such as Hanfu, Qinyang, Huashuo evidently performing well in different regions of apple production.

For the introducing the currently cultivated apple varieties in China to the apple breeders and growers in the world, seven apple breeders and their team members of CARS have taken three years to edit the book of Chinese Apple Varieties, in which 144 varieties were described with beautiful pictures and clarified writings, and as well as double languages. I believe that this book should be a nice reference both for the professionals and the amateurs.

Shu Huairui
June, 2015

前　言

　　苹果是世界上栽培最为普遍的落叶果树之一，最初主要分布于欧洲、亚洲和北美洲，经过2 000多年的驯化和品种选育，现在五大洲都有苹果栽培。中国是苹果属植物的发源地之一，经过长期的自然选择和人工选育，现已拥有极为丰富的苹果种质资源。

　　近年来，随着我国对苹果产业发展的重视，苹果栽培面积趋于稳定，产量稳步增长，栽培区域逐步集中，品种结构有所改善，产业化水平不断提高。从分布来看，我国苹果生产主要集中在环渤海湾、西北黄土高原、黄河故道和西南冷凉高地四大产区。据农业部统计，2013年我国苹果栽培面积和产量分别达到227.2万hm²和3 968.3万t，居世界首位。近年来，我国苹果新品种选育成绩斐然，相继育成了以秦阳、华硕、寒富等为代表的300余个新品种，在我国苹果品种的更新换代和结构调整中起到了重要作用。目前，虽然红富士占据了苹果种植的主要地位，从国外引进的一些新品种，如乔纳金、嘎拉系、藤牧1号、美八等也有一定的发展。

　　为促进我国苹果产业的发展，让国内外从事苹果育种研究的同行对我国目前的栽培苹果品种有所了解，国家苹果产业技术体系育种与资源利用研究室各专家历时3年，采用一致的拍摄标准，对144个苹果品种进行了拍摄，并组织编写了《中国苹果品种》。每个品种分别记述了来源或选育单位、主要形态特征和生物学特性以及品质特性。本书比较全面地反映了中国目前的苹果栽培品种类型以及近年来的选育成就，可以为国内外苹果科研、育种及生产方面相关人员提供一定的参考。

　　本书编写过程中，得到了国家苹果产业技术体系各位同行的诸多支持，在此一并致谢。由于编者业务水平和经验有限，收集的资料不够全面，书稿整理过程中难免存在疏漏，敬请读者批评指正。

编著者

2015年6月

FOREWORD

Apple is one of the most popular deciduous fruit trees cultured in the world. It is originally distributed in Europe, Asia and North America. With more than 2 000 years of natural domestication and breeding improvement, apple has been widely grown in five continents of the world. China is one of the origin places for *Malus* species and has abundant apple germplasm resources through long-term natural selection and artificial breeding.

In recent years, great attention has been paid to the development of apple industry in China. Apple growing areas tend to be stabile and growing regions are gradually concentrated. Total annual production is increased steadily, variety compositions are continuously optimized, and the level of industrialization is continuously increased as well. Apple production in China is mainly distributed in four regions, that is, Bohai Gulf Area, Northwest Loess Plateau, Ancient Cannel Region of Yellow River and Southwest Cold Highland. According to the statistics of China Ministry of Agriculture, the apple production area reached 2.272 million hectars, with a total annual production of 39.683 million tons in 2013, therefore, China is the top country of apple production in the world. Great achievements have been made in apple breeding in China and more than 300 new varieties and strains have been selected such as Qinyang, Huashuo and Hanfu, in last several decades, playing an important role in upgrading apple varieties. Although Fuji is still the dominating variety in China at present, some new varieties introduced from abroad also has got a certain development, such as Jonagold, Gala, MATO, Meiba and so on.

In order to promote the development of apple industry and give a chance to the worldwide apple breeders to know apple varieties cultivated in China, Chinese apple breeders from Breeding and Resource Division of China Apple Industrial Technology System have taken three years to edit and publish the book Apple Varieties in China, in which 144 apple varieties have been described with photos, origin, breeding organization, main morphological characters, biological characteristics and quality traits of each variety using unified standard. This book in fact is a brief introduction of the current apple varieties and breeding achievements in China. It of course could be a good reference for the people working in apple industry at home and abroad.

We are very grateful for receiving a lot of support and assistance from the colleagues of China Apple Industrial Technology System during the course of compiling the book. Criticisms and rectifications from the readers are much appreciated for the shortcomings of the book due to our limited experience.

Editors
June, 2015

目 录
CONTENTS

第一部分
中国苹果概况

Part Ⅰ   General Situation of Apple Industry in China

一、中国苹果栽培历史悠久

A Long History of Apple Cultivation in China

中国是世界上苹果栽培历史最悠久的国家之一，苹果在中国至少有2 200多年的栽培历史。中国的栽培苹果被命名为 *Malus × pumila* Mill.，可分为两大类：中国苹果和西洋苹果。中国苹果古称"奈"，起源于中国新疆西部；汉武帝时代的《上林赋》（公元前126—前118年）已有关于"奈"的记载。稍后，《西京杂记》也提到"奈三：白奈、紫奈、绿奈"。西洋苹果是指目前各国所栽培的苹果，品种大多出自欧美，1871年开始传入中国。无论是中国苹果还是西洋苹果都存在一个漫长的演变过程，都是经过久远的自然选择和人工选择的结果。

China is one of the countries with the longest history of apple cultivation in the world; China has a more than 2 200-year long history of apple production. The cultivated apple of China was named *Malus × pumila* Mill., including *Malus domestica* subsp. *chinensis* Li Y. N. and *Malus domestica* Borkh. The *M. domestica* subsp. *chinensis* Li Y. N. was named 'Nai' in ancient China, and originated in western Xinjiang. It was recorded in *Shang Lin Fu* in the Western Han Dynasty era (from 126 B.C. to 118 B.C.). The article *Xi jing za ji* also recorded three kinds of 'Nai' including white, purple, and green ones. The *M. domestica* Borkh mainly refers to the major apple cultivars of most countries, and the varieties are mostly from Europe and the United States, in 1871 began to spread to China. Both *M. domestica* subsp. *chinensis* Li Y. N. and *M. domestica* Borkh existed a long evolution process, were all the results of a long natural selection and artificial selection.

二、中国苹果种质资源丰富

China Is Rich in Apple Germplasm

中国是世界苹果属植物的大基因中心。世界上苹果属野生种为27种，原产我国的苹果野生种有16种，为世界的59%。我国的云南、贵州及四川三省密集分布15种苹果野生种，包括台湾林檎、尖嘴林檎及滇池海棠等古老的野生种，形成了苹果属植物遗传多样性中心。另一个多样性中心则是新疆天山山脉，此处形成了新疆野苹果大的自然群落。中国不仅苹果野生种质资源丰富，而且苹果栽培历史悠久，拥有遗传多样极为丰富的苹果种质资源。《中国果树志·苹果卷》描述了苹果野生及砧木资源66份，中国原产品种176份，国内选育品种162份，国外引进品种322份。目前，位于辽宁、吉林、云南及新疆的国家果树种质圃保存苹果资源1 750余份（含部分重复），包括野生类型、农家品种、选育品种、遗传材料及国外引进品种。

China is one of the largest gene centers for *Malus* Mill. in the world. There are 27 wild species of *Malus* Mill. in the world, and 59 percent of them which contains 16 wild species originated from China. 15 wild species including some old wild species such as *M. doumeri*, *M. melliana* and *M. yunnanensis* distribute in Yunnan, Guizhou and Sichuan of China. Another diversity center is located in Tian Shan Mountain of Xinjiang where natural communities of *M. sieverii* formed. China is not only rich in wild species of *Malus* Mill., but also has long history of apple cultivation and apple germplasm resources with abundant genetic diversity. 66 wild and rootstock resources, 176 original cultivars of China, 162 domestic bred cultivars and 322 overseas introduced cultivars are described in the book *Chinese Fruit Apple*. At present more than 1 750 apple germplasm resources including some duplicates are preserved in the National Repository of Apple Germplasm Resources of Liaoning, Jilin, Yunnan and Xinjiang, which contains wild types, landraces, bred cultivars, genetic accessions and overseas introduced cultivars.

新疆野苹果林 [*M. sieverii* (Led.) Roem of Xinjiang]

内蒙古山定子 [*M. baccata* (L.) Borkh. of Inner Mongolia]

河北八棱海棠 [*M. robusta* (Carr.) Rehd. of Hebei Province]

甘肃垂丝海棠 （*M. halliana* Koehne of Gansu Province）

三、中国是苹果生产大国

China Is the Largest Apple Producing Country

　　中国是世界上最大的苹果生产国，面积和产量均居世界首位。近年来，中国苹果总产量均稳定在全球苹果总产量的一半左右。据农业部统计，2013年我国苹果种植面积为227.22万 hm²，仅次于柑橘，占我国果园总面积的18.37%，居第二位；总产量为3 968.26万 t，占全国水果总产量的25.16%，居全国首位（表1）。苹果在农业产业结构调整、增加农民收入、促进地方经济快速发展等方面发挥着越来越重要的作用。

China is the largest apple producer in the world, the area and output are all the first in the world. In recent years, the apple's total output is stable in the world, about half of the total apple production. According to the Ministry of Agriculture statistics, in 2013, the total cultivated area of apple in China was 2 272 200 hectares, which second only to the cultivation area of citrus, accounting for 18.37% of the total fruit area of China. The total production of apple in China was 39.6826 million tons, accounting for 25.16% of the total fruit production of China (Table 1).

Apple plays very important roles in many fields, including the structure adjustment of agricultural industry, increasing income, and promoting the rapid development of local economy and so on.

表1 我国主要水果栽培面积和产量（2013）

Table 1 Main fruit cultivated area and production in China （2013）

水果品种 Fruits	栽培面积（万hm²） Area （Ten thousands hectare）	占总面积比例(%) Percentage	产量（万t） Production （Ten thousands ton）	占总产量比例(%) Percentage
苹果 Apples	227.22	18.37	3 968.26	25.16
柑橘 Citrus	242.22	19.57	3 320.94	21.06
梨 Pears	111.17	8.99	1 730.08	10.97
桃 Peaches	76.59	6.19	1 192.41	7.56
葡萄 Grapes	71.46	5.78	1 155.00	7.32
荔枝 Litchis	54.28	4.39	202.25	1.28
香蕉 Bananas	39.20	3.17	1 207.52	7.66
猕猴桃 Kiwifruits	16.15	1.30	176.58	1.12
菠萝 Pineapples	6.05	0.49	138.64	0.88
其他 Others	392.80	31.75	2 679.58	16.99
水果总面积/产量 Fruit total area or production	1 237.14	100.00	15 771.26	100.00

数据来源：中国农业部。

四、中国苹果栽培区域与品种结构

Planting Distribution and Layout of Apples in China

（一）分布区域广泛　Extensive Distribution

我国有适宜于苹果树生长发育理想的地理、土壤和气候条件，苹果分布区域较为广泛，包括渤海湾产区（辽宁、山东、河北等省份）、黄土高原产区（山西、甘肃、陕西、河南等省份）、黄河故道（豫东、鲁西南、苏北和皖北）和秦岭北麓产区（渭河两岸、豫西、湖北西北部），以及西南冷凉高地产区（四川、云南、贵州等省份）、东北寒地小苹果产区、新疆苹果产区。其中，渤海湾产区、黄土高原产区、黄河故道和秦岭北麓产区、西南冷凉高地产区为2008年农业部规划的四大苹果优势主产区。

近年来，中国苹果生产的区域布局发生了较为明显的变化，四大主产区逐步调整为渤海湾和黄土高原两大优势主产区格局，在区域内部也呈现出"西移"和"北扩"趋势。

In China, we have ideal geographical soil and climate conditions, which is suitable for the apple tree growth and development. The distribution area of apple is widely, including Bohai Bay region (Liaoning, Shandong, Hebei, etc), the Loess Plateau region (Shanxi, Gansu, Shaanxi, Henan, etc), the Yellow River communist-held (east of Henan, southwest of Shandong, north of Jiangsu and north of Anhui) and North Qinling region, Southwest cold highland districts (Sichuan, Yunnan, Guizhou, etc), the Northeast cold little apple production region, and Xinjiang apple production region. Among them, the Bohai Bay region, Loess Plateau region, Yellow River communist-held and North Qinling region, and Southwest cold highland districts, were planned as four major advantages regions of producing apples by the Ministry of Agriculture in 2008.

In recent years, the apple production regions layout taken placed obvious changes, the four major advantages regions gradual adjustment for two major producing regions pattern, including Bohai Bay and the Loess Plateau, and within the region also presents "devastated" and "north expansion" trend.

（二）中国苹果品种结构　Variety Structure of China Apple

由于诸多原因，具有悠久栽培历史的原产中国苹果品种现存较少。目前中国苹果生产上主栽品种多为国外引进，红富士占据了苹果种植的主要地位，从国外引进的一些品种，如藤牧1号、美八、乔纳金、嘎拉系等品种也有一定的发展。近年来，我国的苹果新品种选育也有一定突破，育成的寒富、秦阳、华硕等多个新品种在各大苹果产区均有不俗的表现。中国苹果主栽品种与砧木见表2。

Due to many reasons, the existing apple varieties originated from China which has long cultivation history is less. At present, most of cultivation varieties in China are imported, the Fuji occupied the main position of apple planting, some other introduced varieties also had a certain development, including MATO, Meiba, Jonagold and Galas as well. Recently, the apple varieties breeding also have some breakthrough in China, multiple new varieties had good performance in some big apple production areas, such as Hanfu, Qinyang, and Huashuo as well.

表2　中国苹果主要栽培品种与砧木
Table 2　Main cultivars and rootstocks of apple in China

省（自治区、直辖市） Province（autonomous region and municipality）	主栽品种 Main cultivars	主要砧木　Main rootstocks	
		实生砧木 Seedling rootstock	矮化砧木 Dwarfing rootstock
陕西 Shaanxi	富士系、元帅系、嘎拉系、金冠、华冠、秦冠、千秋、乔纳金、秦阳等 Fujis, Delicious, Galas, Golden Delicious, Huaguan, Qinguan, Senshu, Jonagold, Qinyang	楸子 *Malus prunifolia* (Willd.) Borkh	M系，SH系， M series, SH series
山东 Shandong	富士系、元帅系、嘎拉系、金冠、乔纳金 Fujis, Delicious, Galas, Golden Delicious, Jonagold	平邑甜茶、八棱海棠 *M. hupehensis* var. *pinyiensis* Jiang, *M. robusta*(Carr.) Rehd.	M9，M26，SH系，T337 M9, M26, SH series, T337

（续）

省（自治区、直辖市） Province（autonomous region and municipality）	主栽品种 Main cultivars	主要砧木　Main rootstocks	
		实生砧木 Seedling rootstock	矮化砧木 Dwarfing rootstock
河北 Hebei	富士系、元帅系、嘎拉系、金冠、国光、红玉、印度等 Fujis, Delicious, Galas, Golden Delicious, Ralls, Jonathan, Indo	山定子、海棠 *M. baccata* (L.) Borkh., *Malus* sp.	M系，SH系 M series, SH series
甘肃 Gansu	富士系、元帅系、嘎拉系、金冠、秦冠、澳洲青苹等 Fujis, Delicious, Galas, Golden Delicious, Qinguan, Granny Smith	楸子 *M. prunifolia* (Willd.) Borkh.	M系，SH系 M series, SH series
河南 Henan	富士系、元帅系、嘎拉系、秦冠、华冠、金冠、美八、华硕等 Fujis, Delicious, Galas, Qinguan, Huaguan, Golden Delicious, Meiba, Huashuo	河南海棠、八棱海棠 *M. henanhensis* (Pamp.) Rehd., *M. robusta*(Carr.) Rehd.	M系 M series
山西 Shanxi	富士系、元帅系、嘎拉系、秦冠、华冠、金冠、乔纳金、国光等 Fujis, Delicious, Galas, Qinguan, Huaguan, Golden Delicious, Jonagold , Ralls	山定子、湖北海棠、楸子 *M. baccata* (L.) Borkh., *M. hupehensis* (Pamp.) Rehd., *M. prunifolia* (Willd.) Borkh.	M系，SH系，GM256 M series, SH series, GM256
辽宁 Liaoning	富士系、寒富、元帅系、嘎拉系、国光、金冠、乔纳金、华红等 Fujis, Hanfu, Delicious, Galas, Ralls, Golden Delicious, Jonagold , Huahong	山定子 *M. baccata* (L.) Borkh.	GM256，辽砧2号，7734 GM256, LG2, 7734
新疆 Xinjiang	富士系、元帅系、嘎拉系、金冠、寒富等 Fujis, Delicious, Galas, Golden Delicious, Hanfu	塞威氏海棠（新疆野苹果） *M. sieverii* (Led.) Roem.	GM256，M系，SH系 GM256, M series, SH series
云南 Yunnan	富士系、嘎拉系、金冠、华硕等 Fujis, Galas, Golden Delicious, Huashuo	海棠 *Malus* sp.	M系，SH系 M series, SH series
四川 Sichuan	富士系、嘎拉系、金冠、华硕等 Fujis, Galas, Golden Delicious, Huashuo	海棠 *Malus* sp.	M系，SH系 M series, SH series
其他省份（黑龙江、吉林、内蒙古）Other provinces（Heilongjiang, Jilin, Inner Mongolia）	金红、龙冠、龙丰、七月鲜等 Jinhong, Longguan, Longfeng, Qiyuexian	山定子 *M. baccata* (L.) Borkh.	GM256，辽砧2号，7734 GM256, LG2, 7734

一、国内选育品种
Domestic Breeding Varieties

1. 昌红

来源 河北省农林科学院昌黎果树研究所从岩富10全株变异选育的中晚熟品种。2002年通过河北省林木品种审定委员会审定。

主要性状 果实近圆形，平均单果重271g；盖色鲜红色，果面光滑，果点稍大；果肉淡黄色，风味甜酸；可溶性固形物含量17.5%，可滴定酸含量0.38%，果肉硬度8.2kg/cm²；肉质细，汁液多，为优良的鲜食品种。果实发育期145d左右。昌红品种树体适应性强，抗早期落叶病。

1. Changhong

Origin   Changhong is a middle late apple variety selected from tree mutation of Yanfu 10 by Changli Fruit Institute of Hebei Academy of Agriculture and Forestry Sciences. It was examined and approved by Hebei Crop Cultivar Registration Committee in 2002.

Main Characters   The fruit shape is near globose. Its average fruit weight is about 271g. The cover color is bright red, and the fruit surface is smooth with slightly big dots. The flesh is light yellow color with sweet and sour flavor. The content of soluble solid is 17.5%, and the content of total titratable acid is 0.38%. The flesh hardness is 8.2kg/cm². The fruit is crisp, juicy, tasteful, qualitative, and suitable for fresh fruit. The fruit maturation is about 145 days. The cultivar has high adaptability and strong resistance to early defoliation diseases.

2.岱绿

来源 山东省果树研究所从金冠自然实生选育的早熟品种，1989年鉴定并命名。

主要性状 果实短圆锥形，平均单果重166g；果面淡绿色，光滑洁净，无锈；果肉淡黄白色，肉质中粗，汁液多；可溶性固形物含量11.8%，可滴定酸含量0.13%，果肉硬度7.8kg/cm²；风味酸甜，品质上等。果实发育期90～100d。果实较耐贮藏，常温下存放20d，不易皱皮，风味较浓。

2. Dailü

Origin   Dailü is an early apple variety selected from Golden Delicious open pollination by Shandong Fruit Research Institute. It was examined and approved in 1989.

Main Characters   The fruit shape is truncate conical. Its average fruit weight is 166g. The skin of fruit is light green with clean and smooth surface. The flesh is light yellow white color, medium coarse and juicy. The soluble solid content is 11.8%, the total titratable acid 0.13%, the firmness 7.8kg/cm². It is sour-sweet flavor and good eating quality. The fruit developing period is about 90～100 days. It has good storage, can be stored twenty days and no wrinkled skin at room temperature.

3.丹霞

来源 山西省农业科学院果树研究所从金冠实生苗选出的中晚熟品种，原代号72-12-72，1986年通过山西省品种审定。目前在山西等地有栽培。

主要性状 果实圆锥形，平均单果重170.6g；果面底色黄绿，着鲜红色晕，平均着色度75%；果肉乳白色，肉质细脆，汁液多，风味甜；可溶性固形物含量17.0%，总糖含量13.6%，可滴定酸含量0.265%；果肉硬度5.6kg/cm²。果实发育期160d左右。树势中庸，萌芽力中等，成枝力较强；结果早，坐果率高，丰产，采前落果轻。田间表现抗早期落叶病，较抗白粉病；接种鉴定表现高感苹果斑点落叶病和苹果腐烂病，感苹果枝干轮纹病。

3. Danxia

Origin Danxia is a mid-late season cultivar selected from Golden Delicious open pollinated seedlings at Research Institute of Pomology, Shanxi Academy of Agricultural Sciences. The original name of Danxia was 72-12-72 and it was registered in Shanxi province in 1986, and has been commercially used in some places, Shanxi Province.

Main Characters The fruit shape of Danxia is conical and the fruit size is about 170.6g. The fruit ground color is yellowish-green and is covered with splashed Turkey red. The area of cover color is about 75% in average. The flesh is cream-white in color, fine in coarseness, crisp in texture, much juicy and sweet in flavor. The firmness without skin is 5.6kg/cm². The soluble solid content and total titratable acid content are 13.6% and 0.265% respectively. The tree vigor is moderate, the sprouting ratio is high and the shooting ability is strong. Danxia is early bearing and highly productive. The fruit setting ratio is high and the abscission before fruit harvesting is absent. The number of days of fruit growth is about 150. The field resistance to leaf blotchs is high and is relatively high to powdery mildew, but Danxia is highly susceptible to apple *Valsa* canker and *Alternaria* leaf blotch and is susceptible to *Botryosphaeria* canker evaluated by inoculation with the pathogens.

4. 伏翠

来源　中国农业科学院郑州果树研究所以赤阳 × 金冠杂交培育而成的早熟品种，1981年通过鉴定并推广。

主要性状　果实短圆锥形；平均单果重143g左右；果面黄绿色，较光滑；果肉绿白色，肉质松脆，汁液多；风味甜微酸，品质上等；可溶性固形物含量12.9%，可滴定酸含量0.13% ～ 0.20%。果实发育期90d左右，室温条件下果实可贮放15 ～ 20d。

4. Fucui

Origin　Fucui is an early ripening apple variety selected by Zhengzhou Fruit Research Institute, Chinese Academy of Agricultural Sciences. Its parentage is Rainier × Golden Delicious. It was examined and approved in 1981.

Main Characters　The fruit shape is truncate conical. Its average fruit weight is about 143g. The skin color is yellow green with smooth surface. The flesh is green white color, crispy, juicy and sweet with a bit sour. The soluble solid content is 12.9%, the total titratable acid is 0.13%~0.20%. The fruit developing period is about 90 days. The fruit can be stored 15~20 days at room temperature.

5. 伏帅

来源　中国农业科学院郑州果树研究所以长早旭 × 金冠杂交培育而成，1977年命名发表。

主要性状　果实圆锥或长圆锥形；平均单果重140g左右；果皮呈绿黄色，果面光滑，无果锈；果肉淡黄白色，肉质细脆、汁液中多；可溶性固形物含量12.8%，可滴定酸含量0.13%；味甜微酸，有香味，品质上等。果实发育期90d左右，在室温条件下可贮放20d，为早熟品种中较耐贮运的品种。

5. Fushuai

Origin　Fushuai is selected by Zhengzhou Fruit Research Institute, Chinese Academy of Agricultural Sciences. Its parentage is Early McIntosh × Golden Delicious. It was named and published in 1977.

Main Characters　The fruit shape is conical or narrow conical. Its average fruit weight is about 140g. The skin color is green yellow with smooth surface and without fruit rust. The flesh is yellow white color, fine, crispy and juicy. The soluble solid content is 12.8%, the total titratable acid 0.13%. Its eating quality is good with sweet and a little bit sour flavor and aromatic. The fruit developing period is about 90 days. The fruit can be stored 20 days at room temperature. It is one of the more suitable for storage and transportation variety in the early mature varieties.

6.福丽

来源　青岛农业大学以特拉蒙（Telamon）×富士（Fuji）杂交选育的苹果新品种。2014年通过山东省科技厅组织的技术鉴定。

主要性状　果实近圆形，平均单果重239.8g；果面光洁，未套袋果实全面着浓红色；汁液中多，风味甘甜，香气浓郁，可溶性固形物含量16.7%（对照富士15.2%），可滴定酸含量0.28%（富士0.29%），果肉硬度9.5kg/cm²；品质佳，10月中旬成熟。果实极耐贮藏，无需套袋栽培。

6. Fuli

Origin　Fuli is an excellent fresh-eating variety selected by Qingdao Agricultural University. Its parentage is Telamon × Fuji.

Main Characters　The fruit shape is near globose. Its average fruit weight is about 239.8g. The skin of the fruit is in bright dark red color with no bagging culture. The fruit firmness is 9.5kg/cm², and the soluble solid content 16.7% (Fuji is 15.2%), the titratable acid 0.28% (Fuji is 0.29%). The fruit is tasteful and qualitative with pleasant aroma and has higher storability. The fruits can be easily colored without bagging. The fruits ripen in mid-October in Qingdao. The fruits store very well.

7. 福艳

来源　青岛农业大学以特拉蒙（Telamon）×富士（Fuji）杂交选育的生食晚熟品种，2006年通过山东省林木良种审定。

主要性状　果实近圆形，单果重249g；果面光洁，果实底色黄绿，果面大部着鲜红色；果肉黄白色，肉质细而松脆，可溶性固形物含量14.3%，含糖量12.60%，可滴定酸含量0.21%，果肉硬度7.0kg/cm²；汁液多，味甜，风味浓，香气浓郁，品质极上；在烟台地区果实10月上旬成熟；较抗轮纹病，果实在冷藏条件下可贮2个月。

7. Fuyan

Origin　Fuyan is a late-ripening and fresh-eating variety selected by Qingdao Agricultural University. Its parentage is Telamon × Fuji. It was examined and approved by Shandong Variety Registration Committee in 2006.

Main Characters　The fruit shape is near globose. The average fruit weight is 249g. The skin of the fruit is partially in bright red color. The flesh is in yellow-white color and crispy. The fruit firmness is 7.0kg/cm², the soluble solid content 14.3%, the sugar content 12.60% and the total titratable acid 0.21%. The fruit is juicy, sweet, rich in flavor and aroma. The fruit quality is excellent. The fruits ripen in early October and is resistant to ring rot disease (*Botryosphaeria dothidea*). The storage life of the fruits is about two months in cold store.

8. 福早红

来源　青岛农业大学以特拉蒙（Telamon）×新红星（Starkrimson）杂交选育的生食早熟品种，2006年通过山东省林木良种审定。

主要性状　果实圆锥形，单果重225.5g；果实底色黄绿，果面着浓红色，果面光洁；果肉白，肉质松脆，可溶性固形物12.0%，含糖量10.60%，可滴定酸含量0.15%，果肉硬度9.40kg/cm²；汁液中多，味酸甜，香气浓郁，品质上等。在烟台地区果实8月上、中旬成熟；果实贮藏后易发绵。

8. Fuzaohong

Origin　Fuzaohong is an early ripening and fresh eating variety selected by Qingdao Agricultural University. Its parentage is Telamon × Starkrimson. It was examined and approved by Shandong Crop Variety Registration Committee in 2006.

Main Characters　The fruit shape is conical. The average fruit weight is 225.5g. The skin of fruit is in dark red color. The flesh is in white color, loose, crispy and aromatic. The soluble solid content is 12.0%, the sugar content 10.60%, the total titratable acid 0.15%, the firmness 9.40kg/cm², in good balance of sugar and acid. The quality is very good. The fruits mature in early-mid August and is easy to soften after storage.

9. 寒富

来源 沈阳农业大学园艺系与内蒙古宁城县巴林试验场1978年以东光×富士杂交育成的晚熟抗寒苹果新品种。1994年通过内蒙古自治区品种审定，1997年通过辽宁省农作物品种审定委员会审定并命名。

主要性状 果实短圆锥形；平均单果重250g；果实底色黄绿，阳面片红，可全面着色；果面光滑，果点小；果肉淡黄色，甜酸味浓，有香气；可溶性固形物含量15.2%，可滴定酸含量0.34%，果肉硬度9.9kg/cm²；肉质酥脆多汁，为鲜食品种。果实发育期150d左右。丰产性好，树冠紧凑，矮生性状明显。抗寒性与适应能力优于国光和富士，抗苹果粗皮病，较抗蚜虫和早期落叶病。

9. Hanfu

Origin   Hanfu is a new apple variety selected by Horticulture Department of Shenyang Agriculture University and Balin station, Guningcheng City, Inner Mongolia Autonomous Region. It is late ripening and good resistance to hardness variety. Its parentage is Dongguang × Fuji. It was examined and approved by Inner Mongolia Autonomous Region Crop Cultivar Registration Committee in 1994 and by Liaoning Crop Cultivar Registration Committee in 1997.

Main Characters   The fruit shape is short conical. Its average fruit weight is about 250g. The fruit ground color is yellowish-green. The skin of fruit is in full red color. The fruit surface is smooth with small dots. The flesh is in light yellow color and in strong taste of sweet and sour. The soluble solid substance content is 15.2%, the total titratable acid is 0.34%, the firmness is 9.9kg/cm^2. Hanfu is productive, Crown is compact. The tree is dwarf obviously. The flesh is crispy and juicy. It is suitable for fresh eating. The fruit developing period is about 150 days. It is excelled to Ralls and Fuji in hardness and adaptability. It is resistant to branch ring rot disease, aphides and early defoliation disease.

10. 华脆

来源 中国农业科学院果树研究所以金冠 × 惠杂交育成的中晚熟新品种。2010年通过辽宁省农作物品种审定委员会审定。

主要性状 果实长圆形或圆锥形；平均单果重203g；盖色红色，着色有条纹；果面光洁，果粉较少，果点稀而小，外观漂亮；果肉黄白色，风味酸甜适度；可溶性固形物含量为12.8%，可滴定酸含量0.7%，果肉硬度8.1kg/cm²；肉质松脆，汁多味浓，制汁性状优良，出汁率79.9%，果汁基本不褐变。果实发育期160d左右。丰产性好。华脆果实抗果实轮纹病、斑点落叶病能力均强于金冠，抗寒性与金冠相当。

10. Huacui

Origin Huacui is a new apple variety selected by Research Institute of Pomology, Chinese Academy of Agricultural Sciences. Its parentage is Golden Delicious × Megumi. It was examined and approved by Liaoning Crop Cultivar Registration Committee in 2010.

Main Characters The fruit shape is conical or oblong globose. Its average fruit weight is about 203 g. The skin of fruit is red color with bright red skin, good skin finish and small dots. The flesh is in yellow white color and in good balance of sugar and acids. The soluble solid substance content is 12.8%, the total titratable acid 0.7%, the firmness 8.1kg/cm². Huacui is very productive, crispy, juicy, tasteful and qualitative. It is suited for making juice, the juice yield is 79.9%. The fresh juice is seldom brown. The fruit developing period is about 160 days. Compared with Golden Delicious, it has similar tolerant to cold weather and better resistant to fruit ring rot disease and early defoliation disease.

11. 华丹

来源　中国农业科学院郑州果树研究所以美八 × 麦艳杂交培育而成，2013年通过河南省林木品种审定。

主要性状　果实圆锥形，平均单果重160g；果实底色黄绿，果面着鲜红色；果面平滑，蜡质多，有光泽，果点小；果肉白色，肉质中细，松脆，汁液中多；可溶性固形物含量12.4%，可滴定酸含量0.42%，果肉硬度6.3kg/cm^2；风味酸甜，品质中上。果实发育期75～80d，结果性状好、丰产，为早熟鲜食品种。

11. Huadan

Origin　Huadan is a new apple variety selected by Zhengzhou Fruit Research Institute, Chinese Academy of Agricultural Sciences. Its parentage is Meiba × Maiyan. It was examined and approved by Henan Forest Cultivar Registration Committee in 2013.

Main Characters　The fruit shape is conical. Its average fruit weight is about 160g. The skin of fruit is yellow-green with red color, smooth and glossy with wax coat and small dots. The flesh is white color, fine textur, crisp and juicy. The soluble solid content is about 12.4%, the total titratable acid 0.42%, the firmness 6.3kg/cm^2. Flavor is sour-sweet giving good quality. The fruit developing period is about 75~80 days. It is an early mature variety for eating with good fruiting habit and productive.

12. 华富

来源　中国农业科学院果树研究所用长富2号的花药进行花药培养育成的晚熟新品种。2004年通过辽宁省农作物品种审定委员会审定。

主要性状　果实近圆形；平均单果重240g；盖色红色，着色有条纹；果面光滑，果点小；果肉淡黄色，风味酸甜适度；可溶性固形物含量为16.5% ~ 18.2%，果肉硬度9.2kg/cm²；肉质硬脆中细，汁多味浓，品质优。果实发育期170d左右。丰产性好。华富品种适宜在富士栽培区域种植。

12. Huafu

Origin　Huafu is a new apple variety selected by Research Institute of Pomology, Chinese Academy of Agricultural Sciences. It is bred from anther cultured. It was examined and approved by Liaoning Crop Cultivar Registration Committee in 2004.

Main Characters　The fruit shape is near globose. Its average fruit weight is about 240g. The skin of fruit is in red color with dark red stripes, good skin finish and small dots. The flesh is in light yellow color and in good balance of sugar and acids. The soluble solid substance content is 16.5%~18.2%, the firmness 9.2kg/cm². Huafu is very productive, crispy, juicy, tasteful and qualitative. The fruit developing period is about 170 days. It is tolerant to cold weather and resistant to branch ring rot disease. Suitable for the regions where Fuji is planted.

13. 华冠

来源　中国农业科学院郑州果树研究所用金冠 × 富士杂交培育的中晚熟品种，1994通过河南省和山西省农作物品种审定，2002年通过全国农作物品种审定。

主要性状　果实圆锥或近圆形，平均单果重180g；底色绿黄，果面着鲜红条纹；果肉淡黄色，肉质致密，脆而多汁，风味酸甜适宜，有香味，可溶性固形物含量14.0%，可滴定酸含量0.21%，果肉硬度10.2kg/cm²；品质上等。果实发育期160～165d，成熟期比富士早2周，果实耐贮藏。

13. Huaguan

Origin　Huaguan is a new apple variety selected by Zhengzhou Fruit Research Institute, Chinese Academy of Agricultural Sciences. Its parentage is Golden Delicious × Fuji. It was examined and approved by Henan Crop Cultivar Registration Committee in 1994, examined and approved by Chinese crop Cultivar Registration Committee in 2002.

Main Characters　The fruit shape is conical or near globose. Its average fruit weight is about 180g. The skin of fruit is green-yellow with red stripe, smooth and glossy with wax coat and small dots. The flesh is light yellow color, fine-textured, firm to crisp and juicy. The soluble solid content is about 14.0%, the total titratable acid 0.21%, the firmness 10.2kg/cm². Flavor is sour-sweet with aroma and giving good quality. The fruit developing period is about 160~165 days. The mature period is earlier two weeks than Fuji. It has good storage ability.

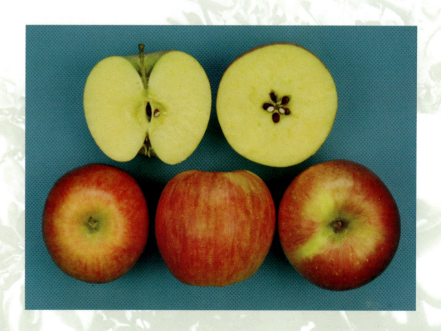

14. 华红

来源　中国农业科学院果树研究所以金冠 × 惠杂交育成的中晚熟新品种。1998年通过辽宁省农作物品种审定委员会审定。

主要性状　果实长圆形；平均单果重250g；盖色红色，着色有条纹；果面光滑，果点小；果肉淡黄色，风味酸甜适度；可溶性固形物含量15.5%，可滴定酸含量0.48%，果肉硬度6.04kg/cm^2；肉质松脆，汁多味浓，品质优，鲜切后不易褐变，为鲜食加工兼用品种。果实发育期150d左右。丰产性好。华红品种树体抗寒性强，抗枝干轮纹病。

14. Huahong

Origin　Huahong is a new apple variety selected by Research Institute of Pomology, Chinese Academy of Agricultural Sciences. Its parentage is Golden Delicious × Megumi. It was examined and approved by Liaoning Crop Cultivar Registration Committee in 1998.

Main Characters   The fruit shape is oblong. Its average fruit weight is about 250g. The skin of fruit is in red color with dark red stripes, good skin finish and small dots. The flesh is in light yellow color and in good balance of sugar and acids. The soluble solid substance content is 15.5%, the total titratable acid 0.48%, the firmness 6.04kg/cm^2. Huahong is very productive, crispy, juicy, tasteful and qualitative. The fresh color can be kept for a long time without browning; therefore, it is suitable for eating and processing. The fruit developing period is about 150 days. It is tolerant to cold weather and resistant to branch ring rot disease.

15. 华金

来源　中国农业科学院果树研究所以金矮生 × 好矮生杂交育成的中晚熟新品种。2004年通过辽宁省农作物品种审定委员会审定。

主要性状　果实圆锥形；平均单果重250g；果皮黄绿色；果面光滑，果点小；果肉白色，风味酸甜适度；可溶性固形物含量15.0%，果肉硬度7.0kg/cm²；肉质松脆，汁多味浓，品质优。果实发育期140d左右。丰产性好。华金品种树体抗寒性强，抗枝干轮纹病。

15. Huajin

Origin　Huajin is a new apple variety selected by Research Institute of Pomology, Chinese Academy of Agricultural Sciences. Its parentage is Gold Spur × Well spur. It was examined and approved by Liaoning Crop Cultivar Registration Committee in 2004.

Main Characters　The fruit shape is conical. Its average fruit weight is about 250g. The skin of fruit is in yellow green color, good skin finish and small dots. The flesh is in light color and in good balance of sugar and acids. The soluble solid substance content is 15.0%, the firmness 7.0kg/cm². Huajin is very productive, crispy, juicy, tasteful and qualitative. The fruit developing period is about 140 days. It is tolerant to cold weather and resistant to branch ring rot disease.

16. 华美

来源 中国农业科学院郑州果树研究所用嘎拉 × 华帅杂交培育而成，2005年通过河南省林木品种审定，2012年通过国家林木品种审定。

主要性状 果实短圆锥形，平均单果重205g；底色淡黄，果面着鲜红色，片状着色；果面光滑，有少量蜡质，果点较大；果肉黄白色，肉质中细，松脆，汁液中多；可溶性固形物含量12.6%，可滴定酸含量0.26%，果肉硬度9.6kg/cm²；风味酸甜适口，有轻微的芳香，品质上等。果实发育期110d左右，比嘎拉早成熟1周左右。

16. Huamei

Origin Huamei is a new apple variety selected by Zhengzhou Fruit Research Institute, Chinese Academy of Agricultural Sciences. Its parentage is Gala × Huashuai. It was examined and approved by Henan Forest Cultivar Registration Committee in 2005, examined and approved by Chinese Forest Cultivar Registration Committee in 2012.

Main Characters The fruit shape is truncate conical. Its average fruit weight is about 205g. The skin of fruit is light yellow with red color, smooth and glossy with wax coat and medium dots. The flesh is yellow white color, fine-textured, crisp and juicy. The soluble solid content is about 12.6%, the total titratable acid 0.26%, the firmness 9.6kg/cm². Flavor is sour-sweet with strong aroma giving a good quality. The fruit developing period is about 110 days. The mature period is earlier one week than Gala.

17. 华瑞

来源　中国农业科学院郑州果树研究所以美八×华冠杂交培育而成，2014年通过河南省林木品种审定委员会审定。

主要性状　果实近圆形，平均单果重220g；底色绿黄，果面着鲜红色，着色面积达60%。果面平滑，蜡质多，有光泽，果点小；果肉黄白色，肉质中细，脆，汁液多；可溶性固形物含量12.8%，可滴定酸含量0.31%，果肉硬度9.7kg/cm^2；风味酸甜适口，浓郁，有芳香；品质上等。果实发育期105d左右，较嘎拉早2周成熟；田间表现抗苹果斑点落叶病。

17. Huarui

Origin　Huarui is a new apple variety selected by Zhengzhou Fruit Research Institute, Chinese Academy of Agricultural Sciences. Its parentage is Meiba × Huaguan. It was examined and approved by Henan Forest Cultivar Registration Committee in 2014.

Main Characters　The fruit shape is near globose. Its average fruit weight is about 220g. The skin of fruit is green-yellow with red stripes (more than 60% surface with red), smooth and glossy with wax coat and small dots. The flesh is yellow-white color, fine textur, crisp and juicy. The soluble solid content is about 12.8%, the total titratable acid 0.31%, the firmness 9.7kg/cm^2. Flavor is sour-sweet and with weak aroma giving good quality. The fruit developing period is about 105 days. The mature period is earlier two weeks than Gala. It is tolerant to apple *Alternaria* leaf spot disease.

18. 华帅

来源 中国农业科学院郑州果树研究所用富士 × 新红星杂交培育而成的新品种，1996年通过河南省农作物品种审定委员会审定。

主要性状 果实圆锥形，平均单果重210g；果面光滑，底色黄绿，果面着暗红色条纹；果面光滑，有果粉，果点中大；果肉黄白色，肉质中粗、松脆，果汁多；可溶性固形物含量13.8%，可滴定酸含量0.26%，果肉硬度7.2kg/cm²；风味酸甜，有芳香，品质上等。果实发育期170d左右。

18. Huashuai

Origin   Huashuai is a new apple variety selected by Zhengzhou Fruit Research Institute, Chinese Academy of Agricultural Sciences. Its parentage is Fuji × Starkrimson Delicious. It was examined and approved by Henan Crop Cultivar Registration Committee in 1996.

Main Characters   The fruit shape is conical. Its average fruit weight is about 210g. The skin of fruit is yellow-green with dark red stripe, smooth and glossy with bloom coat and medium dots. The flesh is yellow-white color, medium-textured, crisp and juicy. The soluble solid content is about 13.8%, the total titratable acid 0.26%, the firmness 7.2kg/cm². Flavor is sour-sweet with aroma and giving good quality. The fruit developing period is about 170 days.

19. 华硕

来源 中国农业科学院郑州果树研究所用美八×华冠杂交培育而成，2009年通过河南省林木品种审定委员会审定，2014年通过国家林木品种审定委员会审定。

主要性状 果实长圆形，平均单果重242g；果实底色绿黄，果面着鲜红色；果面平滑，蜡质多，有光泽；果点小；果肉黄白色；肉质中细，硬脆，汁液中多；可溶性固形物含量13.4%，可滴定酸含量0.31%，果肉硬度10.1kg/cm^2；酸甜适口，风味浓郁，有芳香；品质上等。果实发育期110d左右，比嘎拉早成熟1周，耐贮性好。田间表现较抗苹果轮纹病，对白粉病中度敏感。

19. Huashuo

Origin   Huashuo is a new apple variety selected by Zhengzhou Fruit Research Institute, Chinese Academy of Agricultural Sciences. Its parentage is Meiba × Huaguan. It was examined and approved by Henan Forest Cultivar Registration Committee in 2009. Examined and approved by Chinese Forest Cultivar Registration Committee in 2014.

Main Characters   The fruit shape is oblong. Its average fruit weight is about 242g. The skin of fruit is green-yellow with red color, smooth and glossy with wax coat and small dots. The flesh is yellow white color, fine-textured, firm to crisp and juicy. The soluble solid content is about 13.4%, the total titratable acid 0.31%, the firmness 10.1kg/cm^2. Flavor is sour-sweet and with aroma giving a good quality. The fruit developing period is about 110 days. The mature period is earlier one week than Gala. The fruit have good storage ability, good tolerance to apple ring spot and moderately susceptibility to powdery mildew.

20. 华玉

来源　中国农业科学院郑州果树研究所用藤牧1号 × 嘎拉杂交培育而成，2008年通过河南省林木品种审定委员会审定。

主要性状　果实圆锥形，平均单果重176克；果实底色黄白，果面着鲜红色条纹；果面平滑，有光泽，果粉多；果肉黄白色，肉质细、脆，汁液多；可溶性固形物含量14.2%，可滴定酸含量0.29%，果肉硬度6.2kg/cm²；风味酸甜适口，风味浓郁，有清香；品质上等。果实发育期105～110d，比嘎拉早成熟2周。果实有采前落果现象。

20. Huayu

Origin Huayu is a new apple variety selected by Zhengzhou Fruit Research Institute, Chinese Academy of Agricultural Sciences. Its parentage is MATO × Gala. It was examined and approved by Henan Forest Cultivar Registration Committee in 2008.

Main Characters The fruit shape is conical. Its average fruit weight is about 176g. The skin of fruit is yellow white with red stripe, smooth and glossy with bloom coat and small dots. The flesh is yellow white color, fine-textured, crisp and juicy. The soluble solid content is about 14.2%, the total titratable acid 0.29%, the firmness 6.2kg/cm². Flavor is sour-sweet and palate fullness with strong aroma giving a good quality. The fruit developing period is about 105~110 days. The mature period is earlier two weeks than Gala. Fruits are prone to drop before harvest.

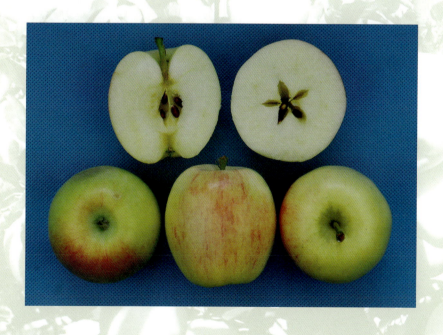

21. 华月

来源　中国农业科学院果树研究所以金冠 × 华富杂交育成的晚熟新品种。2010年通过辽宁省农作物品种审定委员会审定。

主要性状　果实圆柱形；平均单果重200g；果皮黄绿色，阳面有红晕；果面光滑，果点小；果肉淡黄色，风味酸甜适度；可溶性固形物含量15.0%，可滴定酸含量0.46%，果肉硬度7.04kg/cm²；肉质松脆，汁多味浓，品质优。果实发育期170d左右。丰产性好。抗寒性较强，高抗苹果早期落叶病，高抗果实轮纹病。

21. Huayue

Origin　Huayue is a new apple variety selected by Research Institute of Pomology, Chinese Academy of Agricultural Sciences. Its parentage is Golden Delicious × Huafu. It was examined and approved by Liaoning Crop Cultivar Registration Committee in 2010.

Main Characters　The fruit shape is oblong. Its average fruit weight is about 200g. The skin of fruit is in yellow green color with bright red skin, good skin finish and small dots. The flesh is in light yellow color and in good balance of sugar and acids. The soluble solid substance content is 15.0%, the total titratable acid 0.46%, the firmness 7.04kg/cm². Huayue is very productive, crispy, juicy, tasteful and qualitative. The fruit developing period is about 170 days. It is tolerant to cold weather and resistant to branch ring rot disease and early defoliation disease.

22. 鸡冠

来源　起源不详。1926年后在辽宁南部逐渐扩大栽培。

主要性状　果实扁圆形或近圆形，平均单果重160g；盖色鲜红色，并有紫红色断续条纹。果面光滑，果点小；果肉黄白色，风味甜酸；可溶性固形物含量12.4%，可滴定酸含量0.51%，果肉硬度10.1kg/cm²；肉质致密，汁液较多，适于鲜食。果实发育期为160d左右。丰产性强。树体适应性强，较抗旱、抗寒，并对腐烂病抗性强。

22. Jiguan

Origin　The origin of Jiguan is unknown. It was gradually cultivated in southern part of Liaoning Province in 1926.

Main Characters　The fruit shape is oblate or near globose. Its average fruit weight is about 160g. The cover color of skin is turkeyt red with purplish red stripes. The fruit surface is smooth with small dots. The flesh is yellow white with sweet and sour flavor. The content of soluble solid is 12.4%, and the content of total titratable acid is 0.51%. The flesh hardness is 10.1kg/cm². Jiguan is very productive, crisp, juicy, and suitable for fresh fruit. The fruit maturation is about 160 days. The tree has strong vigor and strong resistance to drought, cold and Valsa mali disease.

23. 金红

来源　1960年吉林省农业科学院果树研究所以金冠×红太平杂交育成的中熟苹果品种。

主要性状　果实卵圆形；果个较小，平均单果重75g；果面底色黄色，具鲜红色霞和深红条纹。果点中大，疏。果肉黄白色，味甜酸而浓；可溶性固形物含量12.0%，可滴定酸含量0.6%；肉质松脆，汁液多，有香味，耐贮藏，为鲜食品种。果实发育期130d左右。早果丰产，抗寒性强，抗病虫能力也较强，是一个优良的抗寒品种。

23. Jinhong

Origin　Jinhong is a middle ripening apple variety selected by Jinlin Research Institute of Pomology. Its parentage is Golden Delicious × Hongtaiping in 1960.

Main Characters　The fruit shape is ovoid. It is small. Its average fruit weight is about 75g. The ground color is yellow with dark red stripes. Fruit dots are medium size and sparse. The flesh is yellowish-white color. It tastes strong sweet and sour. The soluble solid substance content is 12.0%, the total titratable acid is 0.6%. Jinhong is early ripening and productive, crispy, juicy, aroma, long-storage. It is suitable for fresh eating. The fruit developing period is about 130 days. It is strong resistant to cold weather and strong resistant to pests and disease. It is a good resistant to hardness variety.

24. 金世纪

来源 西北农林科技大学从皇家嘎拉引进优系中选育的浓红型芽变，2009年通过陕西省农作物品种审定委员会审定。

主要性状 果实圆锥形，高桩，果形指数0.90；果个适中，平均单果重210g；底色黄绿，盖色鲜红色，片红；果面洁净，果点中大，果粉少；风味酸甜，具香气，可溶性固形物含量14.2%，可滴定酸含量0.24%，果肉硬度6.93kg/cm²；果肉黄白色，肉质中细、致密，汁液多；较耐贮藏；果实发育期120d左右。坐果率高，丰产、稳产；对主要病虫害抗性较强。

24. Jinshiji

Origin   Jinshiji is a bud mutation of Royal Gala, selected by Northwest A&F University. It was examined and approved by Shaanxi Crop Cultivar Registration Committee in 2009.

Main Characters   The fruit is conical with a fruit shape index of 0.90, medium in size with an average weight of 210 g. Yellow green ground color with Turkeyt red overcolor, good surface finish, medium-large dots and thin fruit powder. The flavor is aromatic, sweet-tart with a soluble solid substance content of 14.2% and a total titratable acid content of 0.24%. The flesh is yellow white, fine and juicy, with a firmness of 6.93kg/cm². Good storability. The period of fruit development is about 120 days. High fruit setting rate with high and stable yields. High resistance to main diseases and pests.

25. 锦秀红

来源 中国农业科学院郑州果树研究所从华冠中选出的早熟浓红芽变，2009年通过河南省林木品种审定委员会审定。

主要性状 果实圆锥形，平均果重205g；底色绿黄，果面全面着鲜红色，充分成熟后呈浓红色；果面光洁、无锈；果肉淡黄色，肉质细、致密，脆而多汁，风味酸甜适宜；可溶性固形物含量14.2%，可滴定酸含量0.21%，果肉硬度9.9kg/cm^2；品质上等。果实发育期150d左右，熟期与红星接近，果实耐贮、不沙化。

25. Jinxiuhong

Origin Jinxiuhong is a new apple variety selected by Zhengzhou Fruit Research Institute, Chinese Academy of Agricultural Sciences. It is a mutation of Huaguan. It was examined and approved by Henan Forest Cultivar Registration Committee in 2009.

Main Characters The fruit shape is conical. Its average fruit weight is about 205g. The skin of fruit is green-yellow with dark red color, smooth and glossy with wax coat and small dots. The flesh is light yellow color, fine-textured, firm to crisp and juicy. The soluble solid content is about 14.2%, the total titratable acid 0.21%, the firmness 9.9kg/cm^2. Flavor is sour-sweet with aroma and giving good quality. The fruit developing period is about 150 days. The mature period is as same as Red Delicious. It has a good storage ability.

26. 晋富1

来源 山西省果树研究所从红王将中选育的早熟芽变品种。2004年通过山西省品种审定。目前在山西等地有应用。

主要性状 果实近圆形；平均单果重208g；果实底色浅黄，着红色晕；果肉淡黄色，肉质脆，风味甜；可溶性固形物含量15.2%，果肉硬度6.3kg/cm²；果实发育期150d左右。幼树树势强，结果期树势中庸，树姿较开张。树体抗寒优于红王将。

26. Jinfu 1

Origin Jinfu 1 is an early season mutant selected from Beni Osho. It was registered by Shanxi province in 2004 after several years of selection and now has been used at some places like Shanxi Province.

Main Characters The fruit shape of Jinfu 1 is near globose and the fruit size is 208g. The ground color is yellowish-green and is covered with splashed red. The flesh is light yellow in color, crisp in texture and sweet in flavor. The fruit developing period of Jinfu 1 is about 150 days. The soluble solid content is 15.2%. The firmness without skin is 6.3kg/cm². The tree vigor is strong at young stage and becomes moderate when fruiting. The tree posture is fairly open. The cold tolerance is stronger than Beni Osho.

27. 晋富3

来源　山西省果树研究所从长富2号中选育的浓红型芽变品种。1998年发现，为整株变异，2007年通过山西省品种审定。目前在山西等地有应用。

主要性状　果实近圆形；平均单果重233g；底色黄绿，着红色晕；果面光滑，果点小；果肉淡黄色，肉质细脆，汁液多，风味甜；可溶性固形物含量15.3%，糖酸比52：1，果肉硬度8.33kg/cm^2。果实发育期180d左右。树势中庸，丰产性强。

27. Jinfu 3

Origin　Jinfu 3 is a red mutant of Changfu 2 selected at Shanxi Institute of Pomology in 1998. It was registered by Shanxi province in 2007 and now has been used at some places like Shanxi province.

Main Characters　The fruit shape of Jinfu 3 is near globose and the fruit size is 233g. The ground color is yellowish-green and is covered with splashed red. The fruit skin is lubricity. The cuticular dot is small. The flesh is light yellow in color, fine in coarseness, crisp in texture, much in juice and sweet in flavor. The soluble solid content is 15.3% and the sugar-acid ratio is 52：1. The firmness without skin is 8.33kg/cm^2. The number of days of fruit growth is about 180. The tree vigor is moderate. The productivity is high.

28. 葵花

来源　河北省农林科学院昌黎果树研究所以金冠 × 红星杂交育成的晚熟品种。1988年通过河北省农作物品种审定委员会审定。

主要性状　果实扁圆形，平均单果重170g；底色黄，阳面稍有淡紫红晕；果面光滑，果点小；果肉淡黄色，风味甜，微酸；可溶性固形物含量13.6%，可滴定酸含量0.28%，果肉硬度7.0kg/cm²；结果早，丰产，果实无锈，适于鲜食。果实发育期为150d左右。树体长势旺，适应性和抗性比较强。

28. Kuihua

Origin　Kuihua is a late ripening apple variety, which was selected in Changli Fruit Institute of Hebei Academy of Agriculture and Forestry Sciences. Its parentage is Golden Delicious × Starking Delicious. It was examined and approved by Hebei Crop Cultivar Registration Committee in 1988.

Main Characters　The fruit shape is oblate. Its average fruit weight is about 170g. The ground color of skin is yellow with purplish red on the sunward side. The fruit surface is smooth and with small dots. The flesh is light yellow and tartish with sweet flavor. The content of soluble solid is 13.6%, and the content of the total titratable acid is 0.28%. The flesh hardness is 7.0kg/cm². Kuihua is very productive, early fruit, without rust, and suitable for fresh fruit. The fruit maturation is about 150 days. The trees have strong vigor, high adaptability and strong resistance.

29. 礼泉短富

来源　礼泉短富属富士Ⅰ系的短枝型芽变，1985年在礼泉县东庄乡刘家村苹果园发现，1996年通过陕西省农作物品种审定委员会审定并命名。

主要性状　果实圆形或近圆形，果形指数0.88；大果型，平均单果重272g；底色黄绿，盖色为鲜红到浓红，片红；果面光洁，果点中大、稀，无锈；甜酸适中，有香气；果肉黄白色，肉质细脆，汁多；可溶性固形物含量17.4%，可滴定酸含量0.45%，果肉硬度7.59kg/cm²，耐贮藏；在陕西渭北地区果实10月中下旬成熟。易成花，连续结果能力强，丰产、稳产，适于密植栽培。

29. Liquanduanfu

Origin　Liquanduanfu is a spur type sport of Fuji I Line, found at an apple orchard in Liu Village, Dongzhuang Town, Liquan County in 1985. It was examined and approved by Shaanxi Crop Cultivar Registration Committee, and named in 1996.

Main Characters　The fruit is globose with a fruit shape index of 0.88, large in size with an average weight of 272g. Yellow green ground color with blushed red overcolor, good surface finish and medium-large dots. The flavor is aromatic, in good balance of sugar and acid with a soluble solid substance content of 17.4% and a total titratable acid content of 0.45%. The flesh is yellow white, fine, juicy and crispy with a firmness of 7.59kg/cm². Good storability. Easy initiation of floral bud with high and stable yields, suitable for high density planting.

30. 辽伏

来源 辽宁省果树科学研究所以老笃 × 祝光杂交育成的特早熟新品种。1979年通过辽宁省农作物品种审定委员会审定，1990年获国家发明三等奖。

主要性状 果实扁圆形；平均单果重100g；底色翠绿，稍有暗红条纹；果面光滑，果点小；果肉黄白色，风味甜；可溶性固形物含量为11%，可滴定酸含量为0.25%；肉质细脆，汁液多，品质中上等，为鲜食品种。果实发育期60d左右。丰产性好。较耐高温高湿，抗炭疽病、对粗皮病、轮纹病、白粉病有一定的抗性。

30. Liaofu

Origin Liaofu is a new very early apple variety selected by Liaoning Research Institute of Pomology. Its parentage is Lodu × American Summer pearmain. It was examined and approved by Liaoning Crop Cultivar Registration Committee in 1979. It obtained the third grade of National Invention Prize in 1990.

Main Characters The fruit shape is oblate. Its average fruit weight is about 100g. The ground color of fruit is dark green with a little dark red stripes, good skin surface with small dots. The flesh is yellowish-white color and in sweet taste. The soluble solid substance content is 11%, the total titratable acid is 0.25%. 'Liaofu' is good productive, flesh is fine and crispy, juicy. It is suitable for eating and processing. The fruit developing period is about 60 days. It is tolerant to hot and wet weather and resistant to branch ring rot disease. 'Liaofu' is resistant to apple anthracnose, rough bark, ring spot and powdery mildew of apple.

31. 龙丰

来源 黑龙江省农业科学院牡丹江农业科学研究所以金红 × 白龙杂交育成的中熟品种。1990年通过黑龙江省农作物品种委员会审定。

主要性状 果实扁圆形，平均单果重35g，全面紫红色；果面光滑，果点小而稀；果肉黄白色，贮后变粉红色，风味甜酸；可溶性固形物含量12%，可滴定酸含量1.1%，果肉硬度10.3kg/cm²，可贮至元旦。肉质细脆、汁液中多，品质中上。含果胶2.45%，加工成糖水罐头，色泽鲜艳，酸甜适口，品质可与铃铛果罐头比美。果实发育期120d左右。丰产性好。树体抗寒力较强，较抗病虫。

31. Longfeng

Origin   Longfeng is a new middle ripening apple variety selected by Mudanjiang Agriculture Research Institute of Heilongjiang Agriculture Academy of Sciences. Its parentage is Jinhong × Bailong. It was examined and approved by Heilongjiang Crop Cultivar Registration Committee in 1990.

Main Characters   The fruit shape is oblate. Its average fruit weight is 35g. The skin of fruit is in full purplish color. The fruit surface is smooth with small and sparse dots. The flesh is in yellowish-white color and in pink color after storage. The taste is sweet and sour. The soluble solid substance content is 12%, the total titratable acid is 1.1%, the firmness is 10.3kg/cm². It can storage to New Year's day. Longguan is productive. Flesh is fine and crisp, medium juicy and middle quality. The pectin content is 2.45%. Processed into canned are bright color and sour taste, good taste. The quality of its can is equal to the can of Lingdangguo. The fruit developing period is about 120 days. It is strong resistant to cold weather. It is resistant to disease and pests.

32. 龙冠

来源　黑龙江省农业科学院牡丹江农业科学研究所以金冠×七月鲜杂交育成的抗寒中熟新品种。1987年通过黑龙江省农作物品种审定委员会审定。

主要性状　果实长圆锥形；平均单果重94.5g；底色绿黄，着鲜红霞至全红；果面光滑，果点小；果肉黄白色，风味甜；可溶性固形物含量13%，可滴定酸含量0.17%，果肉硬度5.5kg/cm²；汁液中多，品质中上，不耐贮藏，贮至10月初，风味开始变淡。果实发育期110d左右。丰产性好。树体抗寒性强，较抗黑星病、褐斑病和蚜虫。

32. Longguan

Origin　Longguan is a new middle ripening apple variety selected by Mudanjiang Agriculture Research Institute of Heilongjiang Agriculture Academy of Sciences. Its parentage is Golden Delicious × Qiyuexian. It was examined and approved by Heilongjiang Crop Cultivar Registration Committee in 1987.

Main Characters　The fruit shape is narrow conical. Its average fruit weight is about 94.5g. The ground of fruit is in greenish-yellow color. The skin of fruit is in full red color with fresh red stripes. The fruit surface is smooth with small dots. The flesh is in yellowish-white color and in sweet taste. The soluble solid substance content is 13%, the total titratable acid is 0.17%, the firmness is 5.5kg/cm². Longguan is productive, medium juicy and quality, short storage. Its taste turns light from October. The fruit developing period is about 110 days. It is strong resistant to cold weather. It is resistant to scab, brown spot and aphides.

33. 龙红蜜

来源　山东省龙口市果树研究所从烟红蜜中选出的浓红型、中晚熟、着色系芽变品种，2009年通过山东省农作物品种审定委员会审定。

主要性状　果实圆形，果形指数0.88；单果重180g左右；果面光亮，果皮蜡质层厚，满红，着浓紫红色；果肉乳黄色，果汁中等，硬脆，甘甜，有芳香味，可溶性固形物含量15.0%，果肉硬度7.7kg/cm²；果实发育期150d左右，在烟台地区9月下旬成熟。

33. Longhongmi

Origin　Longhongmi is a thick red, mid-late maturing variety selected from a red sport of Yanhongmi by Institute of Pomology of Longkou. It was examined and approved by Shandong Crop Variety Registration Committee in 2009.

Main Characters　The fruit shape is globose. The average fruit weight is 180g, the fruit shape index 0.88, the soluble solid substance content 15.0% and the firmness 7.7kg/cm². The fruit surface is covered with thick wax. The fruit is in full and thick purple-red color with good skin finish. The flesh is in dark yellow color. The fruits are crispy, sweet, aromatic and moderately juicy. The fruit developing period is about 150 days. The fruits ripen in late September in Yantai.

34. 鲁加2号

来源 青岛农业大学以特拉蒙（Telamon）× 富士（Fuji）杂交选育的中早熟加工品种。2009年通过山东省农作物品种审定委员会审定。

主要性状 果实近圆形，果形指数0.83；平均单果重140g；果面光洁，底色黄绿，阳面有红晕和明显的鲜红条纹；果肉黄白色，肉质松而稍粗，汁液多，不易褐变，风味浓酸，可溶性固形物含量12.0%，总糖9.6%；出汁率75.2%，果实原汁酸度0.45%，浓缩汁酸度3.20%。果实发育期115d，在烟台地区8月下旬成熟。果实易感轮纹病，需适时采收。

34. Lujia 2

Origin Lujia 2 is a new apple variety for juice processing selected by Qingdao Agricultural University. Its parentage is Telamon × Fuji. It was examined and approved by Shandong Variety Registration Committee in 2009.

Main Characters The fruit shape is near globose, and the fruit shape index is 0.83. The average fruit weight is about 140 g. The basal color of the fruit is yellow green with bright red stripes and blush. The flesh is in yellow-white color, juicy, loose and slightly rough. The flesh does not darken immediately after cutting. The soluble solid substance content is 12.0%, the total sugar content 9.6%, the juice rate 75.2%, the juice acidity 0.45%, the concentrated juice acidity 3.20%. The fruit developing period is about 115 days. The fruits ripen in late August in Yantai. The fruits are susceptible to ring rot disease and need to harvest in time.

35. 鲁加5号

来源　青岛农业大学以特拉蒙（Telamon）× 富士（Fuji）杂交选育的加工品种。2005年通过了山东省林木品种审定，2006年获得了国家农业植物新品种保护权。

主要性状　果实近圆柱形，单果重177.6g；果面绿色，着红晕；果肉绿白，肉质疏松稍粗，汁液中，风味特酸，果实原汁酸度0.81%，浓缩汁酸度4.50%，果实原汁和浓缩汁澄清、稳定性好，不易褐变，是果汁加工的优良品种。在烟台果实9月下旬成熟。

35. Lujia 5

Origin　Lujia 5 is a new columnar apple variety for juice processing selected by Qingdao Agricultural University. Its parentage is Telamon × Fuji. It was examined and approved by Shandong Crop Variety Registration Committee in 2005. It was patented by the National Office of New Plant Variety Protection of China in 2006.

Main Characters　The fruit shape is nearly oblong and the average fruit weight is about 177.6g. The skin of fruit is blushed on green base towards the sunny side. The flesh is in green-white color, juicy, loose and slightly rough and very sour. The juice acidity is 0.81%, the concentrated juice acidity 4.50%. The fruit juice is clear and the concentrated juice has good stability. The fruits ripen in late September in Yantai.

36. 绿帅

来源 辽宁省果树科学研究所从金冠实生中选育的中早熟新品种。2003年通过辽宁省农作物品种审定委员会审定。

主要性状 果实圆锥形；平均单果重245g；盖色黄绿色；果面平滑光洁，果点大、稀；果肉黄白色，风味甜酸、爽口；可溶性固形物含量12.8%，可滴定酸含量0.34%，果肉硬度8.1kg/cm²；肉质松脆，汁液多，品质上等，室温可存放15d。果实发育期95d左右。丰产性好。树体抗寒性强，较抗苹果腐烂病、粗皮病。

36. Lüshuai

Origin   Lüshuai is a new apple variety selected from seedling of Golden Delicous by Liaoning Research Institute of Pomology. It was examined and approved by Liaoning Crop Cultivar Registration Committee in 2003.

Main Characters   The fruit shape is conical. Its average fruit weight is about 245g. The skin of fruit is yellow green and smooth with big and sparse dots. The flesh is in yellow white. The flesh is in good balance of sugar and acids. The soluble solid substance content is 12.8%, the total titratable acid is 0.34%, the firmness is 8.1kg/cm². Lüshuai is very productive, crispy, juicy, good quality. It is store for 15 days in nature. The fruit developing period is about 95 days. It is tolerant to cold weather and resistant to apple canker and ring rot disease.

37. 苹帅

来源 河北省农林科学院昌黎果树研究所以向阳红 × 胜利杂交育成，属中晚熟品种，2009年通过河北省林木品种审定委员会审定。

主要性状 果实近圆形；平均单果重203g；盖色浓红；果面光滑，蜡质明显；果肉黄白色或浅黄色，质细，硬脆，多汁，香气浓，甜酸适口；可溶性固形物含量14.6%，果肉硬度8.7kg/cm²；品质优。果实发育期150d左右。树体较抗枝干轮纹病，耐瘠薄，适应性强。

37. Pingshuai

Origin   Pingshuai is a new apple variety selected by Changli Institute of Pomology, Hebei Academy of Agriculture and Forestry Sciences. Its parentage is Xiangyanghong × Victory. It was examined and approved by Hebei Crop Cultivar Registration Committee in 2009.

Main Characters   The fruit shape is near globose. Its average fruit weight is about 203g. The skin of fruit is strong red color, good skin finish. The flesh is in light yellow color and in good balance of sugar and acids, crispy, juicy. The soluble solid substance content is 14.6%, the firmness 8.7kg/cm². Pingshuai is very qualitative. The fruit developing period is about 150 days. It is tolerant to barren and resistant to branch ring rot disease.

38. 苹艳

来源　河北省农林科学院昌黎果树研究所以岩富10 × 红津轻杂交育成的中晚熟品种，2009年通过河北省林木品种审定委员会审定。

主要性状　果实近圆形；平均单果重172g；盖色鲜红色，有条纹；果肉浅黄，酸甜适口；可溶性固形物含量15.8%，可滴定酸含量0.20%，果肉硬度8.5kg/cm²；肉质松脆，汁液中多，风味浓甜。果实发育期150d左右。丰产性好。具有抗病性强、耐瘠薄等特点。

38. Pingyan

Origin　Pingyan is a new apple variety selected by Changli Institute of Pomology, Hebei Academy of Agriculture and Forestry Sciences. Its parentage is Yanfu 10 × Red Tsugarau. It was examined and approved by Hebei Crop Cultivar Registration Committee in 2009.

Main Characters　The fruit shape is near globose. Its average fruit weight is about 172g. The skin of fruit is red color with dark red stripes. The flesh is in light yellow color and in good balance of sugar and acids. The soluble solid substance content is 15.8%, the total titratable acid 0.20%, the firmness 8.5kg/cm². Pingyan is very productive, crispy, juicy, tasteful and qualitative. The fruit developing period is about 150 days. It is tolerant to barren and resistant to disease.

39. 七月鲜

来源　辽宁省果树科学研究所以佚名大苹果 × 铃铛果杂交育成的早熟新品种。1958年命名，2006年通过辽宁省非主要农作物品种备案办公室备案。

主要性状　果实卵圆形；平均单果重50.7g；果实颜色鲜艳，果面光滑，富光泽，外观美；果肉黄白色，风味甜酸，有香气；可溶性固形物含量13.94%，可滴定酸含量1.22%，果肉硬度10.99kg/cm^2；肉质脆，汁液多，分期成熟，不耐贮藏，为鲜食、加工兼用品种。果实发育期86d左右。丰产性好。该品种树体抗寒、抗病能力强，对中国北方寒冷地区的自然条件有很强的适应能力。

39. Qiyuexian

Origin   Qiyuexian is a new early ripening apple hybrid selected from unknown cultivar × Dolgo Crab (Lingdangguo) by Liaoning Research Institute of Pomology. It was named in 1958. It was examined and approved by Liaoning Crop Cultivar Registration Committee in 2006.

Main Characters   The fruit shape is ovate globose. Its average fruit weight is 50.7 g. The fruit skin color is in fresh red color. The fruit surface is smooth with small dots. The flesh is in yellowish-white color and in taste of sour and sweet, aroma. The soluble solid substance content is 13.94%, the total titratable acid is 1.22%, the firmness is 10.99kg/cm^2. It is productive. The flesh is crispy, juicy. It is ripening in different stage. It is short storage. It is suitable for fresh eating and processing. The fruit developing period is about 86 days. It is strong cold and disease resistance. It is strong adaptability to cold nature condition in north area of China.

40. 秦富1号

来源 西北农林科技大学于20世纪90年代初在周至县从长富2号上发现的富士短枝型芽变优系，在陕西渭北多地试栽表现良好。

主要性状 果实近圆形，高桩；大果型，平均单果重300g；底色黄白，盖色浓红色，片红；果面光洁，果点中、稀；风味酸甜，具香气，可溶性固形物含量16.3%，可滴定酸含量0.35%，果肉硬度7.44kg/cm^2；果肉黄白色，肉质细脆、汁多；耐贮藏。果实10月中下旬成熟。易形成短枝，树形紧凑，适于密植栽培。

40. Qinfu 1

Origin   Qinfu 1 is a late season apple variety, a spur type sport of Changfu 2, selected by Northwest A&F University in Zhouzhi County in the early 1990s. It shows good adaptability in Weibei Area of Shaanxi Province.

Main Characters   The fruit is near globose, large in size with an average weight of 300g. Cream-white ground color with blushed strong red overcolor, good surface finish and medium-large dots. The flavor is aromatic, sweet-tart with a soluble solid substance content of 16.3% and a total titratable acid content of 0.35%. The flesh is yellow white, fine, juicy and crispy with a firmness of 7.44kg/cm^2. Good habit, suitable for high destiny planting.

41. 秦冠

来源　原陕西省果树研究所以金冠×鸡冠杂交选育的晚熟品种，1957年杂交，1970年定名，是目前我国栽培面积较大的自育品种之一。

主要性状　果实圆锥形或短圆锥形，果形端正；大果型，平均单果重250g；底色黄绿，盖色紫红色，着红色条纹，充分成熟时全面着色；果点大；风味甜，可溶性固形物含量16.0%，可滴定酸含量0.26%，果肉硬度8.84kg/cm²；有香气；果肉黄白色，肉质较粗、硬韧，汁中多；耐贮运；果实10月中下旬成熟。易成花，早果性好，连续结果能力强，丰产、稳产；适应性强，抗旱、抗寒、抗病性较强。

41. Qinguan

Origin　Qinguan is a late season apple variety, originated from the cross Golden Delicious × Jiguan made by Research Institute of Pomology, Shaanxi Province. It was named in 1970, and it is widely cultivated throughout China.

Main Characters　The fruit is conical or Truncate-conical, large in size with an average weight of 250 g. Yellow green ground color with strong red overcolor and large dots. The flavor is aromatic, sweet with an soluble solid substance of 16.0% and an total titratable acid content of 0.26%. The flesh is yellow white, coarse, juicy and firm with a firmness of 8.84kg/cm². Good storability. Precocious fruit bearing with high and stable yields. Strong adaptability, drought and cold tolerance and disease resistance.

42. 秦红

来源　西北农林科技大学以威赛克旭×嘎拉杂交选育的中晚熟品种，1991年杂交，2011年通过陕西省农作物品种审定委员会审定。

主要性状　果实长圆锥形，果形指数0.95，果形端正、高桩；大果型，平均单果重265g；盖色红色，着条红；果面光洁，果点小、中多；风味甜酸，果肉黄白色，肉质松脆、稍粗，汁多；可溶性固形物含量14.6%，可滴定酸含量0.63%，果肉硬度7.3kg/cm^2，较耐贮藏。果实9月上中旬成熟，果实发育期150d左右。易形成短枝，连续结果能力强，坐果率高，丰产、稳产。对主要病虫害抗性较强。

42. Qinhong

Origin　Qinhong is a mid-late season apple variety, originated from the cross Wijcik × Gala made by Northwest A&F University in 1991. It was examined and approved by Shaanxi Crop Cultivar Registration Committee in 2011.

Main Characters　The fruit is narrow conical with a fruit shape index of 0.95, large in size with an average weight of 265g. Striped red overcolor with good surface finish and fine dots. The flavor is sweet-tart with a soluble solid substance content of 14.6% and a total titratable acid content of 0.63%. The flesh is yellow white, medium-coarse, juicy and crispy with a firmness of 7.3kg/cm^2. Good storability. It blooms in mid-late April, and matures in early September. The fruit developing period is about 150 days. Spur type, high fruit setting rate with high and stable yields. Resistance to main diseases and pests.

43. 秦星

来源   原陕西省果树研究所以新红星×秦冠杂交选育的中晚熟品种，1975年杂交，1995年通过陕西省农作物品种审定委员会审定。

主要性状   果实近圆柱形；大果型，平均单果重302g；底色淡绿，盖色浓红色，全面着色；果面光洁，无锈，果点小，蜡质多；酸甜适口，有香味；果肉黄白色，质地松脆，汁多；可溶性固形物含量14.4%，可滴定酸含量0.17%，果肉硬度7.85kg/cm^2；耐贮藏，在室内常温下贮藏3周左右。在陕西渭北南部9月中旬成熟。具短枝特性，早果、丰产。适应性广，对苹果主要病虫害抗性较强。

43. Qinxing

Origin   Qinxing is a mid-late season apple variety, originated from the cross Strakrimson Delicious × Qinguan made by Research Institute of Pomology, Shaanxi Province in 1975. It was examined and approved by Shaanxi Crop Cultivar Registration Committee in 1995.

Main Characters   The fruit is near globose, large in size with an average weight of 302g. Light green ground color with completed strong red overcolor, good surface finish, small dots and thick waxiness. The flavor is aromatic, in good balance of sugar and acid with a soluble solid substance content of 14.4% and a total titratable acid content of 0.17%. The flesh is yellow white, juicy and crispy with a firmness of 7.85kg/cm^2. Good storability at room temperature for up to 3 weeks. Spur type, high productive with precocious fruit bearing. Good adaptability and highly resistance to main diseases and pests.

44. 秦艳

来源 原陕西省果树研究所以天王红×秦冠杂交选育的中晚熟品种，1975年杂交，2000年通过陕西省农作物品种审定委员会审定。

主要性状 果实圆锥形，果形指数0.85；平均单果重242g；底色黄绿，盖色浓红色，全面着色；果面光洁，无锈，蜡质厚；风味酸甜，香味较浓；果肉黄白色，肉质松脆，多汁；可溶性固形物含量14.4%，可滴定酸含量0.41%，果肉硬度7.39kg/cm^2，耐贮藏。果实9月中下旬成熟，果实发育期150～160d。抗性强，适应性广。

44. Qinyan

Origin   Qinyan is a mid-late season apple variety, originated from the cross Tianwanghong × Qinguan made by Research Institute of Pomology, Shaanxi Province in 1975. It was examined and approved by Shaanxi Crop Cultivar Registration Committee in 2000.

Main Characters   The fruit is conical with a fruit shape index of 0.85 and an average weight of 242g. Yellow green ground color with completed strong red overcolor, good surface finish and thick waxiness. The flavor is intensely aromatic, sweet-tart with a soluble solid substance content of 14.4% and a total titratable acid content of 0.41%. The flesh is yellow white, juicy and crispy with a firmness of 7.39kg/cm^2. The fruit developing period is 150~160 days. Good resistance and adaptability.

45. 秦阳

来源 西北农林科技大学从皇家嘎拉自然杂种后代中选育的早熟品种，2005年通过陕西省农作物品种审定委员会审定。

主要性状 果实近圆形，果形指数0.86，果形端正；平均单果重198g；底色黄绿，盖色鲜红色，着红色条纹，充分成熟时全面呈鲜红色；果面光洁，无锈，果点白色，中大、中多；风味甜，可溶性固形物含量12.2%，可滴定酸含量0.38%，果肉硬度8.32kg/cm²；有香气；果肉黄白色，肉质细脆，汁液中多；室温条件下可贮藏10～15d。果实7月中下旬成熟。早果、丰产性好，抗病性较强。

45. Qinyang

Origin   Qinyang is a new early season apple variety, originated from the open pollinated seedlings of Royal Gala by Research Institute of Pomology in College of Horticulture, Northwest A&F University. It was examined and approved by Shaanxi Crop Cultivar Registration Committee in 2005.

Main Characters   The fruit is near globose with a fruit shape index of 0.86 and an average weight of 198g. Yellow green ground color with striped orange red overcolor, good surface finish and medium-large white dots. The flavor is aromatic, sweet with a soluble solid substance of 12.2% and a total titratable acid content of 0.38%. The flesh is yellow white, fine, juicy and crispy with a firmness of 8.32kg/cm². Storage life at room temperature for 10~15 days. Precocious fruit bearing with high productive. Disease resistance.

46. 秋富红

来源　辽宁省果树科学研究所从秋富1苹果中选育出的短枝型芽变新品种。2010年通过辽宁省非主要农作物品种备案办公室备案。

主要性状　果实长圆形；平均单果重280g；盖色浓红色，底色黄绿；果面光滑，果点中大；果肉淡黄色，风味甜酸适口；可溶性固形物含量15.2%，可滴定酸含量0.47%，果肉硬度10.8kg/cm^2；肉质松脆，汁液多，品质优，为鲜食品种。果实发育期170d左右。树冠矮小紧凑，短枝率高。树体抗寒性略强于红富士，较抗枝干轮纹病。

46. Qiufuhong

Origin　Qiufuhong apple, a spur-type bud mutant, was selected from Qiufu 1 by Liaoning Research Institute of Pomology. It was examined and approved by Liaoning Crop Cultivar Registration Committee in 2010.

Main Characters　The fruit shape is oblong. Its average fruit weight is about 280g. The skin of fruit is in dark red color, the ground color is in yellow green color. Fruit surface is smooth with middle dots. Flesh is in light yellow color and in good balance of sugar and acids. The soluble solid substance content is 15.2%, the total titratable acid is 0.47%, the firmness is 10.8kg/cm^2. The flesh is crispy, juicy and qualitative. It is suitable for fresh eating. The fruit developing period is about 170 days. Qiufuhong crown is dwarf and compact, spur rate is high. It is little more tolerant to cold weather than red Fuji and resistant to branch ring rot disease.

47. 秋锦

来源 中国农业科学院果树研究所以国光×（红冠+元帅+旭+金冠+倭锦）杂交育成的晚熟品种。1988年通过辽宁省农作物品种审定委员会审定。

主要性状 果实近圆形；平均单果重175g；盖色紫红色；果面光滑，果点中大；果肉淡黄色，风味甘甜；可溶性固形物含量16.0%，可滴定酸含量0.20%，果肉硬度9.8kg/cm²；肉质松脆，汁液中多，风味浓甜。果实发育期160d左右。丰产性好。

47. Qiujin

Origin Qiujin is a new apple variety selected by Research Institute of Pomology, Chimese Academy of Agricultural Sciences. Its parentage is Ralls Janet × (Richard Delicious +Golden Delicious+McIntosh+Red Delicious). It was examined and approved by Liaoning Crop Cultivar Registration Committee in 1988.

Main Characters The fruit shape is near globose. Its average fruit weight is about 175g. The skin of fruit is purplish red color, good skin finish and medium large dots. The flesh is in light yellow color and in good balance of sugar and acids. The soluble solid substance content is 16.0%, the total titratable acid 0.20%, the firmness 9.8kg/cm². Qiujin is very productive, crispy, juicy, tasteful and qualitative. The fruit developing period is about 160 days.

48. 秋口红

来源　山东省果树研究所于1972年在海阳县徐家店镇姜家秋口村果园发现的自然实生种。2009年通过山东省林木良种审定委员会审定。

主要性状　果实圆形或扁圆形；单果重180g；果实底色黄绿，阳面着鲜红色霞，具红条纹；果肉黄白，肉质酥脆而稍粗，果汁中多，风味香甜适口，可溶性固形物含量14.9%，可滴定酸含量0.44%，每100g果肉维生素C含量2.94g，果肉硬度8.99kg/cm^2。果实发育期120d左右，8月中旬成熟；抗旱，耐瘠薄，较抗轮纹病，易感霉心病，有采前落果现象。

48. Qiukouhong

Origin　Qiukouhong is a variety selected from an occasional seedling, found in the village of Jiangjia Qiukou of Xujiadian Town of Haiyang County by Shandong Institute of Pomology in 1972. It was examined and approved by Shandong Trees Variety Registration Committee in 2009.

Main Characters　The fruit shape is globose or oblate. Its average fruit weight is about 180g. The skin of fruit is brightly red striped on yellow green base. The flesh is in yellow-white color, crispy, juicy, tasteful and in good balance of sugar and acids. The soluble solid substance content is 14.9%, the total titratable acid 0.44%, the Vc content 2.94 g/100g, the firmness 8.99kg/cm^2. The fruits ripen in mid-August, the fruit developing period is about 120 days. It is tolerant to arid, barren and branch ring rot disease, it's easily susceptible to the mould core disease, and has fruits drop phenomenon before harvest.

49. 陕富6号

来源　西北农林科技大学在2000年从富士Ⅰ系中选出的浓红型芽变优系，在陕西渭北多地试栽表现良好。

主要性状　果实圆形或近圆形，果形指数0.86；平均单果重280g；底色黄绿或淡黄，盖色鲜红色，片红；果面光洁，无锈，果点中、稀；酸甜适中，可溶性固形物含量16.5%，可滴定酸含量0.36%，果肉硬度8.14kg/cm²；有香味；果肉黄白色，肉质致密、细脆，汁多。果实10月中下旬成熟。

49. Shanfu 6

Origin　Shanfu 6 is a late season maturity, a bud mutation variety of Fuji I Line, selected by Northwest A&F University in 2000. It shows good adaptability in Weibei Area of Shaanxi Province.

Main Characters　The fruit is globose with a fruit shape index of 0.86 and an average weight of 272g. Yellow green or light yellow ground color with Turkeyt red overcolor, good surface finish and medium-large dots. The flavor is aromatic, in good balance of sugar and acid with a soluble solid substance content of 16.5% and a total titratable acid content of 0.36%. The flesh is yellow white, fine, juicy and crispy with a firmness of 8.14kg/cm².

50. 沈农2号

来源　沈阳农学院选育而成，亲本不详，杂交种1951年取自原兴城园艺试验场。

主要性状　果实圆锥形，平均单果重80g；底色黄绿，呈浓红色霞；果面平滑，果点小；果肉乳白色，风味甜酸；可溶性固形物含量13%；肉质细而致密，汁液多，微有芳香，品质中等，耐贮性稍差。果实发期120d左右。较丰产。抗寒力较强。

50. Shennong 2

Origin　Shennong 2 is a new apple variety selected by Shenyang Agriculture College. Its parentage is unknown. The crossed seeds were from former Xingcheng Horticulture Station in 1951.

Main Characters　The fruit shape is conical. Its average fruit weight is about 80 g. The fruit ground color is yellowish-green with skin color in great red blush. The fruit surface is smooth with small dots. The flesh is in milky color and in taste of sweet and sour. The soluble solid substance content is 13%. Shennong 2 is very productive. Flesh is fine and close, juicy, a little aroma and middle qualitative. It is a little weak in storage. The fruit developing period is about 120 days. It is strong tolerant to cold weather.

51. 胜利

来源 河北省昌黎果树研究所以青香蕉 × 倭锦杂交育成的晚熟品种。1988年通过河北省农作物品种审定委员会审定。

主要性状 果实短圆锥形，平均单果重170g；盖色暗红色，并有断续红条纹；果面较粗糙，果点大；果肉浅黄色，风味甘甜；可溶性固形物含量14.9%，可滴定酸含量0.21%，果肉硬度10.6kg/cm²；肉质硬脆，汁液多，可适于鲜食。果实发育期为160d左右。丰产性强。树体长势旺，适应性强，但果面易生果锈。

51. Shengli

Origin Shengli is a late-maturing apple variety bred by White Pearmain × Ben Davis in Changli Fruit Institute of Hebei Academy of Agriculture and Forestry Sciences. It was examined and approved by Hebei Crop Cultivar Registration Committee in 1988.

Main Characters The fruit shape is truncate conical. Its average fruit weight is about 170g. The cover color of skin is dark red with dark red stripes.The fruit surface is rough with big dots. The flesh is light yellow with sweet flavor. The content of soluble solid is 14.9%, and the content of total titratable acid is 0.21%.The flesh hardness is 10.6kg/cm². Shengli is very productive, crisp, juicy, and suitable for fresh fruit.The fruit maturation is about 160 days. The cultivar has strong vigor and high adaptability, but sometimes there are fruit russeting on the fruit surface.

52. 双阳红

来源　青岛农业大学以特拉蒙（Telamon）×（嘎拉+Falstaff+新世界）杂交选育的早中熟苹果新品种。2014年通过山东省农作物品种审定委员会审定。

主要性状　果实近圆形，果形指数0.86，平均单果重153.2g；果实外观光洁，果形端庄，果面红色；果肉黄白色，果肉脆，酸甜爽口，可溶性固形物含量15.1%，果肉硬度7.97kg/cm^2；香气浓郁，品质上等。在青岛地区9月上旬成熟。不需套袋栽培。

52. Shuangyanghong

Origin　Shuangyanghong is a fresh-eating variety selected by Qingdao Agricultural University. Its parentage is Telamon × (Gala+Falstaff+Shinsekai). It was examined and approved by Shandong Crop Variety Registration Committee in 2014.

Main Characters　The fruit shape is near globose, and the shape index is 0.86. The average fruit weight is about 153.2g. The skin of the fruit is in bright red color. The flesh is in yellow-white color and crispy. The fruit firmness is 7.97kg/cm^2, the soluble solid content 15.1%. The fruit is qualitative with pleasant aroma. The fruits can be easily colored without bagging. The fruits ripen in early September in Qingdao.

53. 泰山嘎拉

来源 山东省果树研究所从皇家嘎拉中选出的早中熟芽变品种。2010年通过山东省农作物品种审定委员会审定。

主要性状 果实圆锥形，果形指数0.84，平均单果重212.8g；果面着色鲜红，底色黄绿，全面着片红，果面光滑；果心小，果肉淡黄色，肉质细、硬脆，汁液多，甜酸适度，有香气。可溶性固形物含量15.0%，可溶性糖含量13.8%，可滴定酸含量0.39%，果肉硬度为7.2kg/cm²。早果性和丰产性好，抗病性强。在泰安地区8月10日左右果实成熟。

53. Taishan Gala

Origin   Taishan Gala is a new early-medium-ripening variety selected from bud mutation of Royal Gala by Shandong Institute of Pomology. It was examined and approved by Shandong Crop Variety Registration Committee in 2010.

Main Characters   The fruit shape is conical, and the fruit shape index is 0.84. The average fruit weight is about 212.8g. The skin is brightly red blushed on yellow-green base and covered. Its core is small and flesh is in light yellow color and finely textured, crispy and juicy with moderate sweet-sour flavor and pleasant aroma. The fruit firmness is 7.2kg/cm², the soluble solids content 15.0%, the soluble sugar content 13.8%, the total titratable acid 0.39%. The quality is excellent. It is a high yield variety with strong resistance to diseases. The fruits ripen in early August in Taian.

54. 泰山早霞

来源　山东农业大学从苹果实生苗中选出的早熟品种。2008年通过山东省农作物品种审定委员会审定。

主要性状　果实短圆锥形，果形指数0.93；平均单果重138.6 g；果面光洁，底色淡黄，色相条红；果肉白色，可溶性固形物含量12.8%，糖酸比21.2∶1，酸甜适口，有香气。果实发育期70～75d，在泰安地区6月25日前后成熟。

54. Taishanzaoxia

Origin　Taishanzaoxia is an early-ripening variety selected from apple seedling by Shandong Agricultural University. It was examined and approved by Shandong Crop Variety Registration Committee in 2008.

Main Characters　The fruit shape is truncated conical. Its fruit shape index is 0.93, the average fruit weight 138.6 g, the soluble solid substance content 12.8%, the ratio of sugar and acid 21.2. The smooth skin of fruit is striped on light yellow base. The flesh is in white color and juicy with moderate sweet-sour flavor and pleasant aroma. The fruit developing period is about 70~75 days. The fruits mature in 25th, June in Taian area.

55. 天汪1号

来源 天水市果树研究所选育的红星短枝型芽变品种，1980年在天水市泰州区杏湾村果园发现，1999年通过甘肃省林木良种审定委员会审定，2003年通过国家林业局林木品种审定委员会审定。

主要性状 果实圆锥形，果顶五棱明显，果形端正、高桩，果形指数0.92～0.98；果个大，平均单果重180～200g，最大果重365g；底色黄绿，着全面鲜红或浓红色，片红；果面光滑，有光泽，鲜艳美观；风味香甜，可溶性固形物含量11.9%～14.1%，可滴定酸含量0.21%，果肉黄白色，略带绿色，肉质细、致密，汁多。果实9月中旬成熟，发育期141～148d。耐贮性与新红星相近。

55. Tianwang 1

Origin Tianwang 1 is a new mid-season and spur-type apple variety, originated from bud mutation of Starking Delicious. It was discovered at an orchard of Apricot-bay Village in Tianshui in 1980, then approved by Gansu Improved Forestry Varieties Approving Committee and Crop Cultivar Registration Committee of State Forestry Bureau in 1999 and 2003, respectively.

Main Characters The fruit is conical with a fruit shape index 0.92~0.98, large in size with an average weight of 180~200g. Yellow green ground color with complete over blushed bright or dark red and good surface finish. The flavor is aromatic, sweet with a soluble solid substance content of 11.9%~14.1% and a total titratable acid content of 0.21%. The flesh is yellow white with slight green, fine, juicy. The storability is similar with Starkrimson. The period of fruit development is 141~148 days.

56. 望山红

来源　辽宁省果树科学研究所从长富2号中选出的早熟、浓红芽变新品种。2004年通过辽宁省非主要农作物品种备案办公室备案。

主要性状　果实近圆形；平均单果重260g；盖色鲜红色，着色有条纹；果面光滑，果点中大；果肉淡黄色，风味酸甜爽口；可溶性固形物含量15.3%，可滴定酸含量0.38%，果肉硬度9.2kg/cm²；肉质松脆，汁液多，品质上等，为鲜食品种。比长富2号提前15d成熟，果实发育期160d左右。丰产性好。

56. Wangshanhong

Origin　Wangshanhong is a new early ripening and thick red apple variety mutation selected from Changfu 2 by Liaoning Research Institute of Pomology. It was examined and approved by Liaoning Crop Cultivar Registration Committee in 2004.

Main Characters　The fruit shape is oblate. Its average fruit weight is 260g. The skin of fruit is in fresh red color with red stripes. The fruit surface is smooth with middle size dots. The flesh is in light yellow color and in good balance of sugar and acids. The soluble solid substance content is 15.3%, the total titratable acid is 0.38%, the firmness is 9.2kg/cm². Wangshanhong is productive, crispy, juicy and qualitative. It is suitable for fresh eating. It is ripening 15 days earlier than Changfu 2. The fruit developing period is about 160 days.

57. 向阳红

来源　河北省昌黎果树研究所以小国光 × 红星杂交育成的晚熟品种。

主要性状　果实短圆锥形；平均单果重200g；盖色暗红色；果面光滑，果点小；果肉淡黄色，肉质较细；可溶性固形物含量10.1%，果肉硬度5.2kg/cm²；肉质松脆，汁液中多，风味酸甜。果实发育期170d左右。

57. Xiangyanghong

Origin　Xiangyanghong is a new apple variety selected by Changli Institute of Pomology, Hebei Academy of Agriculture and Forestry Sciences. Its parentage is Small Ralls Janet × Starking.

Main Characters　The fruit shape is conical. Its average fruit weight is about 200g. The skin of fruit is red color, good skin finish with small dots. The flesh is in light yellow color and crispy. The soluble solid substance content is 10.1%, the firmness 5.2kg/cm². Xiangyanghong is very crispy, juicy, tasteful. The fruit developing period is about 170days.

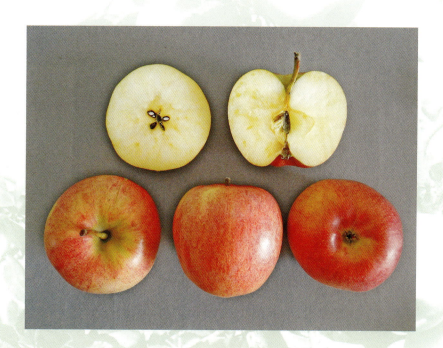

58. 烟富3

来源　烟台市果树工作站从长富2号中选出的富士着色系品种，1997年通过山东省农作物良种评审委员会的审定。

主要性状　果实圆形或长圆形，果形端正，果形指数0.86～0.89；果个大型，平均单果重245g。着色好，片红，色泽浓红艳丽；果实肉质爽脆，汁液多、风味香甜，可溶性固形物含量14.8%～15.4%，果肉硬度8.7～9.7kg/cm^2。品质上等，果实综合性状优于长富2号，外观性状优于长富1号。果实生育期180d，极耐贮藏。

58. Yanfu 3

Origin　Yanfu 3 is a high quality variety selected from red mutation of Changfu 2 Fuji by Fruit and Tea Extension Station of Yantai. It was examined and approved by Shandong Crop Variety Registration Committee in 1997.

Main Characters　The fruit shape is globose to oblong globose. The fruit shape index is 0.86~0.89. The average fruit weight is about 245g. The skin of fruit is in fully bright red color. The flesh is crispy, juicy and sweet. The firmness is 8.7~9.7kg/cm^2. The soluble solid substance content is 14.8%~15.4%. The fruit quality is excellent, better than Changfu 2 and Changfu 1. The fruit developing period is about 180 days. The fruits store very well.

59. 烟富6

来源　烟台市果树工作站从惠民短枝富士中选出的着色良好的短枝型富士品种。1998年通过山东省农作物品种审定。

主要性状　果实扁圆至近圆形，果形指数0.86～0.90；平均单果重253～271g；果面光洁，易着色，色浓红；果肉淡黄色，致密硬脆，汁多，味甜，可溶性固形物含量15.2%，果肉硬度9.8kg/cm²。成熟期10月中旬。果实极耐贮藏。

59. Yanfu 6

Origin　Yanfu 6 is a spur type variety selected from bud mutation of Huimin Spur-type Fuji by Fruit and Tea Extension Station of Yantai. It was examined and approved by Shandong Variety Registration Committee in 1998.

Main Characters　The fruit shape is oblate to near globose, and the shape index is between 0.86 and 0.90. The average fruit weight is about 253~271g. It has good skin finish and is easily colored to dark red. The flesh is in light yellow color, dense, firm, crispy, juicy and sweet. The soluble solid substance content is 15.2% and the firmness is 9.8kg/cm². The fruits ripen in early October. The fruit quality is excellent. The fruits store very well.

60. 烟嘎2号

来源　烟台市果树工作站在蓬莱市湾子口园艺场从嘎拉中选出的着色系品种，1998年通过山东省农作物品种审定委员会审定。

主要性状　果实近圆形或卵圆形，果形指数0.86以上；平均单果重218g；果实底色黄白，果面鲜红；果肉乳黄色，肉质致密，可溶性固形物含量14.4%，果肉硬度6.79kg/cm^2，细脆，汁液多，甜酸适口，品质上等。果实发育日数为110～120d，9月上旬成熟。

60. Yanga 2

Origin　Yanga 2 is a bud mutation of Gala selected in the Wanzikou Horticultural Farm of Penglai county by Pomology and Tea Extension Station of Yantai. It was examined and approved by Shandong Crop Variety Registration Committee in 1998.

Main Characters　The fruit shape of Yanga 2 is near globose or ovoid. The fruit shape index is above 0.86, the average fruit weight about 218g, the soluble solid substance content 14.4%. The skin of fruit is brightly red colored on yellow white base. The finely textured flesh is in milk yellow color and the firmness is 6.79kg/cm^2. Yanga 2 is crispy, juicy, and in good balance of sugar and acids, maturing in early September. The fruit developing period is about 110~120 days.

61. 烟嘎3号

来源　烟台市果树工作站从嘎拉中选出的中早熟、着色系芽变品种。2008年通过山东省农作物品种审定委员会审定。

主要性状　果实近圆至卵圆形，果形指数0.85；平均单果重219g；果面色相片红，大部或全部着鲜红色；果肉乳白色，风味浓郁，肉质细脆爽口，可溶性固形物含量12.2%，果肉硬度6.7kg/cm²。果实发育期110～120d，在烟台地区8月底至9月初成熟。可与富士、新红星等互为授粉树。

61. Yanga 3

Origin　Yanga 3 is a new early mid maturing variety selected from Gala bud mutation by Pomology and Tea Extension Station of Yantai. It was examined and approved by Shandong Crop Variety Registration Committee in 2008.

Main Characters　The apple shape is near globose to ovoid, the shape index 0.85, the average fruit weight 219g, the soluble solid substance content 12.2%, the firmness 6.7kg/cm². The skin of fruit is in partial or full bright red. The flesh is in milk-white color. Its flavor is rich and flesh is crispy. The fruits mature at the end of August to the early September in Yantai and the fruit developing period is about 110~120 days. It has good compatibility of pollination with Fuji and Starkrimson Delicious.

62. 烟红蜜

来源 1975年由山东省黄县芦头乡迟家沟的一株偶然实生树选出。

主要性状 果实长圆柱形或圆形，平均单果重180g；果面底色绿黄，全面着鲜红色晕；果肉淡黄色，肉质硬而脆，味甜无酸，具似香蕉的香气，果汁中等，可溶性固形物含量15.4%，果肉硬度7.26kg/cm^2。在烟台地区果实10月中下旬成熟。适应性强，抗旱，耐瘠薄。

62. Yanhongmi

Origin   Yanhongmi is a variety selected from an occasional apple seedling in Chijiagou village of Lutou town of Huangxian county of Shandong province in 1975.

Main Characters   The fruit shape is narrow oblong or globose. The average weight is 180g. The fruit skin is blushed on yellow-green base. The flesh is in light yellow color, firm and crispy. The fruit firmness is 7.26kg/cm^2 and the soluble solid content 15.4%. The fruit is aromatic like banana and the juice is sweet with little acid. It has strong adaptability and is tolerant to drought and poor soil. The fruits ripen in mid to late October in Yantai.

63. 延风

来源　原陕西省果树研究所以元帅×金冠杂交选育的中晚熟品种，1957年杂交，1970年定名，1987年通过陕西省农作物品种审定委员会审定。

主要性状　果实圆锥形，果顶有明显五棱，似元帅品种；平均单果重142g；底色淡黄，盖色淡红色，着条红；果面少光泽，较粗糙，果点小、稀；风味甜，可溶性固形物含量16.8%，香气浓；果肉淡黄色，肉质细密、松脆，汁液中多。果实10月上旬成熟；耐贮藏。果实易感水心病；抗逆性较强。

63. Yanfeng

Origin　Yanfeng is a mid-late season apple variety, originated from the cross Red Delicious × Golden Delicious made by Research Institute of Pomology, Shaanxi Province in 1975. It was named in 1970, examined and approved by Shaanxi Crop Cultivar Registration Committee in 1987.

Main Characters　The fruit is conical, similar to Red Delicious, with an average weight of 142g. Light yellow ground color with striped red overcolor, coarse surface finish and few small dots. The flavor is intensely aromatic, sweet with a soluble solid substance content of 16.8%. The flesh is light yellow, fine, juicy and crispy. Good storability. High susceptibility to watercore. High resistance to stress.

64. 燕山红

来源 河北省昌黎果树研究所以国光 × 红冠杂交育成的晚熟品种。1993年通过河北省农作物品种审定委员会审定。

主要性状 果实短圆锥形或近圆形，平均单果重215g，盖色红色；果面光滑，果点小；果肉浅黄色或黄白色，风味酸甜适度；可溶性固形物含量16.0%，可滴定酸含量0.49%，果肉硬度9.8kg/cm²；果肉质地细，松脆，有香气，品质上等，适于鲜食。果实发育期165d左右。树体适应性和抗逆性均较强。

64. Yanshanhong

Origin Yanshanhong is a late-maturing apple variety bred by Ralls × Richard Del. In Changli Fruit Institute of Hebei Academy of Agriculture and Forestry Sciences. It was examined and approved by Hebei Crop Cultivar Registration Committee in 1993.

Main Characters The fruit shape is truncate conical or near globose. Its average fruit weight is about 215g. The cover color is red, and the fruit surface is smooth with small dots. The flesh is light yellow or yellow white with moderately sour and sweet flavor. The content of soluble solid is 16.0%, and the content of total titratable acid is 0.49%. The flesh hardness is 9.8kg/cm². The fruit is crispy, juicy, aromatic, tasteful, qualitative and suitable for fresh fruit. The fruit maturation is about 165 days. The cultivar has high adaptability and strong stress tolerance.

65. 玉华早富

来源　陕西省果树良种苗木繁育中心从弘前富士芽变选育的中晚熟品种，2005年通过陕西省农作物品种审定委员会审定。

主要性状　果实圆形或近圆形，果形指数0.88；平均单果重231g；底色黄绿或淡黄，盖色鲜红，着色有条纹；果面光洁，无锈；酸甜适中，可溶性固形物含量14.8%，可滴定酸含量0.36%，果肉硬度6.77kg/cm²；有香味；果点较大；果肉黄白色，肉质细脆，汁多。在陕西渭北地区9月中下旬成熟。坐果率较高，连续结果能力优于晚熟富士。

65. Yuhuazaofu

Origin　Yuhuazaofu Fuju is a mid-late season apple variety, a bud mutation of Hirosaki Fuju, selected by Shaanxi Fruit Trees Breeding Center. It was examined and approved by Shaanxi Crop Cultivar Registration Committee in 2005.

Main Characters　The fruit is near globose with a fruit shape index of 0.88 and an average weight of 231g. Green yellow or light yellow ground color with striped Turkeyt red color, good surface finish and medium-large dots. The flavour is aromatic, in good balance of sugar and acid with a soluble solid substance content of 14.8% and a total titratable acid content of 0.36%. The flesh is yellow white, fine, juicy and crispy with a firmness of 6.77kg/cm². High setting rate with better yield stability than Fuji.

66. 岳苹

来源　辽宁省果树科学研究所以寒富 × 岳帅杂交育成的晚熟新品种。2009年通过辽宁省非主要农作物品种备案办公室备案。

主要性状　果实圆锥形；平均单果重295g；果实鲜红色；果面光洁，果点小；果肉黄白色，风味酸甜，微香；可溶性固形物含量15.3%，可滴定酸含量0.22%，果肉硬度11.2kg/cm²；肉质松脆，汁液多，品质中上，为鲜食品种。果实发育期165d左右。丰产、稳产性好。树体抗寒性强，抗枝干轮纹病。

66. Yueping

Origin Yueping is a new late ripening apple hybrid selected from Hanfu × Yueshuai by Liaoning Research Institute of Pomology. It was examined and approved by Liaoning Crop Cultivar Registration Committee in 2009.

Main Characters The fruit shape is conical. Its average fruit weight is 295g. The fruit skin color is in fresh red color. The fruit surface is smooth with small dots. The flesh is in yellowish-white color and in taste of sour and sweet, a little aroma. The soluble solid substance content is 15.3%, the total titratable acid is 0.22%, the firmness is 11.2kg/cm^2. It is productive and stability, crispy, juicy, middle or good qualitative. It is suitable for fresh eating. The fruit developing period is about 165 days. It is strong cold resistance and resistant to branch ring rot disease.

67. 岳帅

来源 辽宁省果树科学研究所以金冠 × 红星杂交育成的晚熟新品种。1995年通过辽宁省农作物品种审定委员会审定并命名。

主要性状 果实近圆形；平均单果重224g；果面着橘红色霞，覆有鲜红条纹；果面光滑，果点小；果肉黄白色，风味酸甜适口；可溶性固形物含量15.47%，可滴定酸含量0.27%，果肉硬度9.36kg/cm²；肉质细脆，汁液多，有香味，品质上等。10 ～ 12月为最佳食用期。果实发育期155d左右。丰产性好。较抗寒、抗病，其适应力与金冠相近。

67. Yueshuai

Origin   Yueshuai is a new late ripening apple variety selected by Liaoning Research Institute of Pomology. Its parentage is Golden Delicious × Starking. It was examined and approved by Liaoning Crop Cultivar Registration Committee

in 1995.

Main Characters   The fruit shape is near round. Its average fruit weight is 224g. The skin of fruit is in nacarat color with fresh red stripes. The fruit surface is smooth with small dots. The flesh is in yellowish-white color and in good balance of sugar and acids. The soluble solid substance content is 15.47%, the total titratable acid is 0.27%, the firmness is 9.36kg/cm^2. Yueshuai is very productive, crispy, juicy, aroma, good quality. It is suitable for eating from October to December. The fruit developing period is about 155 days. It is tolerant to cold weather and resistant to disease. It is similar to Golden Delicious in adaptability.

68. 岳阳红

来源 辽宁省果树科学研究所以富士×东光杂交育成的中晚熟新品种，2009年通过辽宁省非主要农作物品种备案办公室备案。

主要性状 果实近圆形；平均单果重205g；底色黄绿，全面着鲜红色，色泽艳丽；果面光洁，果点小；果肉淡黄色，风味酸甜爽口；可溶性固形物含量15.2%，可滴定酸含量0.50%，果肉硬度10.1kg/cm²；极易着色，肉质松脆，汁液多，品质优，为鲜食品种。果实发育期145d左右。丰产性好。树体抗寒性较强，较抗枝干轮纹病和苹果腐烂病。

68. Yueyanghong

Origin Yueyanghong is a new middle late ripening apple variety selected by Liaoning Research Institute of Pomology. Its parentage is Fuji × Dongguang. It was examined and approved by Liaoning Crop Cultivar Registration Committee in 2009.

Main Characters The fruit shape is oblate. Its average fruit weight is about 205g. The skin of fruit is in full red color with yellowish-green ground, good surface and small dots. The flesh is in light yellow color and in good balance of sugar and acids. The soluble solid substance content is 15.2%, the total titratable acid is 0.50%, the firmness is 10.1kg/cm². Yueyanghong is very productive, easy coloring, crispy, juicy, good quality. It is suitable for fresh eating. The fruit developing period is about 145 days. It is tolerant to cold weather and resistant to branch ring rot disease and apple canker.

69. 早翠绿

来源　山东省果树研究所以辽伏×岱绿杂交选育而成的早熟品种，2003年通过山东省林木品种审定委员会审定。

主要性状　果实圆形或圆锥形，果形指数0.93，平均单果重151.3g；果面光洁，果皮绿色；果肉乳白色或微带淡黄色，致密而脆，果肉硬度8.59kg/cm²；汁液多，酸甜适口，可溶性固形物含量14.2%，总糖含量12.60%，可滴定酸含量0.24%；具有浓郁芳香味，品质中上等。在泰安地区果实7月中旬成熟。

69. Zaocuilü

Origin　Zaocuilü is a early-ripening variety selected by Shandong Institute of Pomology. Its parentage is Liaofu × Dailü. It was examined and approved by Shandong Trees Variety Registration Committee in 2003.

Main Characters　The fruit shape is globose or conical. The shape index is 0.93 and the average fruit weight is 151.3g. It has the good skin finish and the skin is in full green color. The flesh is in milky white color. The fresh is dense, crispy and rich in flavor. The firmness is 8.59kg/cm². The fruits have rich juice and in good balance of sugar and acids. The quality is good. The soluble solid substance content is 14.2%, the total sugar content 12.60%, the total titratable acid 0.24% .The fruits ripen in the mid-July in Taian.

二、国外选育苹果栽培品种

Foreign Breeding Varieties

70. 阿斯

来源　原产美国，1970年在俄勒冈州发现的俄矮红的枝变，为元帅系第五代品种。1988年引入我国。

主要性状　果实圆锥形，高桩，五棱凸起，果个大；着色早，色泽艳丽、一致，成熟时为紫红色；风味较甜，涩味极轻，可溶性固形物含量14.8%，可滴定酸含量0.30%；果肉乳白色，松脆，多汁。果实生育期130d左右，比其他元帅系芽变品种成熟期早。果实耐贮藏，气调条件下可贮至翌年5～8月。树冠较开张，属半矮化类型。

70. Ace

Origin　Ace, originated from United States, is a mutated variety of Oregon Spur. It was found in Oregon in 1970 and introduced into China in 1988.

Main Characters　The fruit is conical with angels on the fruit top, large in size. Purplish red with earlier coloration. The flavor is light acerbity, sweet with a soluble solid substance content of 14.8% and a total titratable acid content of 0.30%. The flesh is cream, loose, crispy and very juicy. The fruit developing period is about 120 days. Storage life under controlled atmosphere for 7 months.

71. 安娜

来源　原产以色列，是国外亚热带地区试种苹果最易成功的品种之一。中国农业科学院郑州果树研究所1984年由美国引入。

主要性状　果实长圆锥形，高桩，平均单果重155g；果实底色黄绿，果面着淡红晕，果皮光洁无锈；果肉乳白色，肉质细、松脆，汁液多，可溶性固形物含量12.5%，风味酸甜适度，味浓，品质中上等。果实发育期110d左右。低温需求量少，仅为一般品种的1/2，适宜于在低纬度地区栽培。

71. Anna

Origin　Anna is one of the most successful apple varieties tested in oversea subtropical regions which is originated from Israel. It was introduced to China from America by Zhengzhou Fruit Research Institute, Chinese Academy of Agricultural Sciences in 1984.

Main Characters　The fruit shape is narrow conical. Its average fruit weight is 155g. The surface of fruit is smooth and clean, covered with light red blush on the yellow green background color. The flesh is cream color, fine, crisp and juicy. The soluble solid content is 12.5%. The eating quality is medium to good with suitable sour sweet flavor and strong aromatic. The fruit developing period is about 110 days. The low chill requirement is very low. It is about a half that of the normal varieties and suitable to plant in the low latitude regions.

72. 奥查金

来源　美国密苏里州果树试验站用金冠 × Hl291（元帅 × Conrad）杂交培育成的品种，20世纪80年代初由日本引入我国。

主要性状　果实圆形或圆锥形，平均单果重178g；底色黄绿或绿黄，阳面稍有水红晕；果皮平滑，有光泽；果肉淡黄色或黄白色，肉质松脆，汁液多；可溶性固形物含量11.0%，可滴定酸含量0.55%，果肉硬度7.2kg/cm^2；风味酸甜，味较浓，微有香气，品质中上等。果实发育期110～120d，比金冠早熟2～3周。

72. Ozark Gold

Origin　Ozark Gold is selected by American Missouri Fruit Experiment Station. Its parentage is Golden Delicious × HI291 (Red Delicious × Conrad). It was introduced to China from Japan in early 1980s.

Main Characters　The fruit shape is globose or conical. Its average fruit weight is 178g. It is very similar to Golden Delicious, fruit sunny side with some light red blush on the yellow green or green yellow color. The flesh is light yellow or yellow white color, crisp and juicy. The soluble solid content is 11.0%, the total titratable acid 0.55%, the firmness 7.2kg/cm^2. It is sour-sweet flavor with slightly aromatic, medium to good eating quality. The fruit developing period is about 110~120 days. The mature period is earlier 2~3 weeks than Golden Delicious.

73. 澳洲青苹

来源 澳大利亚以自然实生选育的晚熟品种，其亲本不详。

主要性状 果实圆锥形或短圆锥形，平均单果重175g；果面翠绿色；果面平滑，果点大；果肉绿白色，风味酸；可溶性固形物含量11.8%，可滴定酸含量0.57%，果肉硬度9.1kg/cm²；肉质紧密、硬脆，酸味重，为加工及鲜食兼用品种。果实发育期为146d左右。树体适应性较强。

73. Granny Smith

Origin   Granny Smith is a late-maturing apple variety selected from natural seedlings in Australia. Its parentage is unknown.

Main Characters   The fruit shape is conical or truncate conical. Its average fruit weight is about 175g. The cover color is green. The fruit surface is smooth with big dots. The flesh is green white with sour flavor. The content of soluble solid is 11.8%, and the content of total titratable acid is 0.57%. The flesh hardness is 9.1kg/cm². Granny Smith is firm, crisp, sour, and suitable for fresh fruit and processing. The fruit maturation is about 146 days. The tree has strong adaptability.

74. 坂田津轻

来源　日本从津轻芽变中选育的红色中晚熟品种。

主要性状　果实圆形或近圆形，平均单果重220g；底色黄绿，盖色为淡红晕，并有断续红色条纹；果面光滑，果点小；果肉黄白色，风味酸甜适度，微有香气；可溶性固形物含量13.8%，可滴定酸含量0.26%，果肉硬度11.3kg/cm²；外观漂亮，甜脆，汁多，风味独特。果实发育期120d左右。丰产性强。树体适应性广，抗病性强，易管理，但有采前轻微落果现象。

74. Sakata Tsugaru

Origin　Sakata Tsugaru is a middle late apple variety selected from bud mutations of Jinqing in Japan.

Main Characters　The fruit shape is globose or near globose. Its average fruit weight is about 220g. The ground color of skin is yellow green, and the cover color is light red with red stripes. The fruit surface is smooth with small dots. The flesh is yellow white, slightly aromatic and with moderately sweet and sour flavor. The content of soluble solid is 13.8%, and the content of total titratable acid is 0.26%. The flesh hardness is 11.3kg/cm². Sakata Tsugaru is very productive with beautiful appearance and unique flavor, and has sweet, crisp and juicy fruit. The fruit maturation is about 120 days. The cultivar has high adaptability and strong resistance, and can be easily managed. But there is slight fruit drop before harvest.

75. 北斗

来源 日本青森县果树试验场以富士 × 陆奥杂交育成的中熟三倍体品种。1983年引入我国。

主要性状 果实近圆形；平均单果重250g；果面光滑、有光泽，底色黄绿，盖色有红色条纹；果肉黄白色，肉质细、松脆；汁液多，风味酸甜适度，有香味；可溶性固形物含量14.0%，果肉硬度7.5kg/cm²。果实发育期150d左右。由于是三倍体品种，树势壮旺，抗逆性强。

75. Beidou

Origin   Beidou is a new triploid apple variety selected by Chuo gyorui. Its parentage is Fuji × Luao. It was introduced into China in 1983.

Main Characters   The fruit shape is near globose. Its average fruit weight is about 250g. The skin of fruit is red color with dark red stripes. The flesh is in yellow white color, crispy, juicy, and in good balance of sugar and acids. The soluble solid substance content is 14.0%, the firmness 7.5kg/cm². The fruit developing period is about 150 days. Beidou is a triploid varieties, so the tree is vigorous, and has better resistance.

76. 布瑞本

来源 1952 年在新西兰发现的哈密尔顿夫人苹果（Lady Hamilton）和澳洲青苹（Granny Smith）杂交后代的芽变品种。

主要性状 果个中大，果底黄色，果皮颜色橙红至红色，果肉脆，多汁，味甜，有香气。晚熟，耐藏性好，鲜食加工兼用。

76. Braeburn

Origin　Braeburn is a variety discovered from a bud mutation generated from the progeny of Lady Hamilton × Granny Smith in Braeburn Orchard of New Zealand in 1952.

Main Characters　The fruit is mediumin large in size. The skin is from scarlet to dark red in colour, striped, on a golden yellow base, smooth and shiny. Elongated shape. Yellow-cream flesh. Long peduncle. It is very versatile, tasty, can be used in many ways, from raw to baked/cooked. The flesh does not darken immediately after cutting. Very firm, crunchy and juicy, finely textured. Aromatic, with a pleasant tart flavour. The fruits store well.

77. 长富2号

来源 富士苹果的芽变品系之一，1980年农业部从日本引入接穗，分给辽宁省果树科学研究所、山东省烟台果树工作站等单位试栽。

主要性状 果实近圆形，平均单果重250g；果面底色黄绿色，成熟时全面浓红鲜艳；盖色红色、着色有条纹；果面光滑，果点中大；果肉黄白色，风味酸甜适度，稍有芳香；可溶性固形物含量15.5%，可滴定酸含量0.48%，果肉硬度11.47kg/cm^2；肉质松脆致密，贮后仍脆而不变，汁液多，品质上等，为鲜食品种。果实发育期170d左右。丰产性好。树体抗寒性差，易患枝干粗皮轮纹病。

77. Changfu 2

Origin Changfu 2 is one of Fuji mutations, the scions were induced from Japen by Agricultural Ministry in 1980 and distributed to Liaoning Research Institute of Pomology and Shandong Yantai Fruit Tree Station to plant in trial.

Main Characters The fruit shape is oblate, slanting. Its average fruit weight is 250g. The fruit ground color is yellowish-green. The skin of fruit is in full thick fresh red color during ripening. The skin color is red with strips. The fruit surface is smooth with middle size dots. The flesh is in yellowish white color and in good balance of sugar and acids, a little aroma. The soluble solid substance content is 15.5%, the total titratable acid is 0.48%, the firmness is 11.47kg/cm^2. Changfu 2 is productive. The flesh is crispy and close. It is also crispy, juicy and qualitative after storage. It is suitable for fresh eating. The fruit developing period is about 170 days. Changfu 2 is weak in tolerant to cold weather and easily to infect branch ring rot disease.

78. 赤阳

来源　赤阳是美国自然实生选育的晚熟品种，引入我国时间不详。

主要性状　果实圆锥形，平均单果重270g；盖色暗红色，并有断续暗红条纹；果面较粗糙，果点大；果肉黄白色，风味酸甜；可溶性固形物含量12.7%，可滴定酸含量0.5%，果肉硬度9.3kg/cm²；肉质硬脆，汁液较多，香味浓郁，可用于鲜食。果实发育期为175d左右。丰产性强。树体适应性强，较耐寒、耐旱。

78. Rainier

Origin　Rainier is a late-maturing apple variety selected from natural seedlings in America. The time when it was introduced into China is unknown.

Main Characters　The fruit shape is conical. Its average fruit weight is about 270g. The cover color of skin is dark red with dark red stripes. The fruit surface is rough with big dots. The flesh is yellow white with sour and sweet flavor. The content of soluble solid is 12.7%, and the content of total titratable acid is 0.5%. The flesh hardness is 9.3kg/cm². Rainier is very productive with firm and crispy, juicy and strong aromatic fruit, and suitable for fresh fruit. The fruit maturation is about 175 days. The tree has high adaptability and strong tolerance to cold and drought.

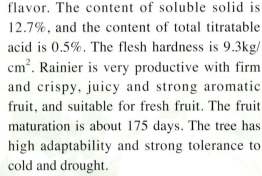

79. 斗南

来源　日本青森县从麻黑7号自然实生苗中选育的品种。1998年10月通过辽宁省农作物品种审定委员会审定。

主要性状　果实圆锥形；平均单果重300g；盖色浓红色；果面光滑，果点大而稀；果肉黄白色，肉质松脆，汁液中多，酸甜，有清香味；可溶性固形物含量16.0%，果肉硬度8.2kg/cm²。果实发育期180d左右。生长势较强，适宜在富士栽培区种植。

79. Dounan

Origin　Dounan is a new apple variety selected from Mahei 7 natural seedlings by Chuo gyorui. It was examined and approved by Liaoning Crop Cultivar Registration Committee in 1998.

Main Characters　The fruit shape is conical. Its average fruit weight is about 300g. The skin of fruit is strong red color, good skin finish and big dots. The flesh is in yellow white color and in good balance of sugar and acids, crispy, juicy. The soluble solid substance content is 16.0%, the firmness 8.2kg/cm². The fruit developing period is about 180 days. The trees of Dounan is vigorous. Suitable for the regions where Fuji is planted.

80. 俄矮2号

来源　原产美国，为俄矮红的芽变，元帅系第五代品种。

主要性状　果实圆锥形，五棱明显，高桩，果形指数0.91～0.95；平均单果重210g，最大果重350～400g；果实着色较早，表色为全面鲜红或浓红，条红；果面光滑，有光泽，鲜艳美观；甜酸适口，有芳香，可溶性固形物含量12%～14%，可滴定酸含量0.26%，品质上等；果肉黄白色，肉质细，汁多。果实发育期135d左右，较耐贮藏。树体半矮化，结果早、丰产。

80. Oregon Spur II

Origin　Oregon Spur II, originated from United States, is a bud mutation variety of Oregon Spur, recognizing as a fifth generation variety of Delicious strain.

Main Characters　The fruit is conical with angels on the fruit top and an average weight of 210g. Full striped strong red over color, good surface finish and glossy. The flavor is aromatic, in good balance of sugar and acid with a soluble solid substance content of 12%~14% and a total titratable acid content of 0.26%. The flesh is light yellow, fine and very juicy. The fruit developing period is about 135 days. High productive with precocious fruit bearing. Semi-dwarf type.

81. 恩派

来源　原产美国，由纽约州农业试验站于1945年以旭和元帅杂交培育而成，1966年发表。20世纪80年代引入我国。

主要性状　果实扁圆形，果顶有五棱；平均单果重150g；果面光滑，有光泽，果粉厚、多；底色黄绿，全面着浓红或暗红色；芳香味浓，甜酸适度，可溶性固形物含量14.4%，可滴定酸含量0.5%，果肉硬度7.7kg/cm²；果肉黄白色，肉质细、松脆，汁多；较耐贮藏。果实发育期约140d。无采前落果，果实病害少。

81. Empire

Origin   Empire is a native to United States apple variety, originated from the cross Golden McIntosh × Red Delicious made by the New York State Agricultural Experiment Station. It was released in 1966 and introduced into China in 1980s.

Main Characters   The fruit is oblate with angels on the fruit top, an average weight of 150g. Yellow green ground color with strong red overcolor, good surface finish and thick fruit powder. The flavor is intense, in good balance of sugar and acid with a soluble solid substance content of 14.4% and a total titratable acid content of 0.5%. The flesh is yellow white, juicy, fine and crispy with a firmness of 7.7kg/cm². The period of fruit development is around 140 days. Good storability. High resistance to diseases.

82. 粉红女士

来源　源自澳大利亚，由西澳洲Stoneville试验站以Lady Williams与金冠杂交育成的极晚熟品种，1979年选出，1985年正式发表。1995年引入我国，2004年陕西省通过品种审定。

主要性状　果实近圆柱形，端正、高桩，果形指数0.94；中果型，平均单果重200g，最大果重306g；底色淡绿，着全面粉红色或鲜红色，色泽艳丽；果面洁净，无锈，果点中大、中密，外观美；果肉乳白色，贮存1～2个月后果肉淡黄色，风味浓郁，甜酸适度，可溶性固形物含量16.65%，可滴定酸含量0.65%，果肉硬度9.16kg/cm²；硬脆，汁中多；极耐贮藏，室温下可贮至翌年4～5月。在陕西渭北南部地区，果实10月下旬至11月上旬成熟，发育期200d左右。早果性好，丰产、稳产，抗病性较强。

82. Pink Lady

Origin　Pink Lady is a new extremely late season apple variety, originated from the cross Golden Delicious and Lady Williams made by the Stoneville Horticultural Research Station near Perth, Western Australia in 1979. It was examined in 1985, introduced into China in 1995 then approved by Shaanxi Fruit Cultivar Registration Committee in 2004.

Main Characters　The fruit is near globose oblong with a fruit shape index of 0.94, medium in size with an average weight of 200g. Light green ground color with complete over pinkish, good surface finish and big dots. The flavour is intense, in good balance of sugar and acid with a soluble solid substance content of 16.65% and a total titratable acid content of 0.65%. The flesh is light yellow, juicy and crisp with a firmness of 9.16kg/cm². Long storage life for up to April even May in next year at room temperature. The fruit development period is about 200 days. High and stable yields. High resistance to virus diseases.

83. 福岛短枝

来源   普通红富士苹果的优良芽变品种，1984年从日本引入我国，1993年11月通过辽宁省农作物品种审定委员会的认定推广。

主要性状   果实近圆形；平均单果重250g；果实片红，色浓；果面光滑，果点小；果肉黄白色，肉质脆而致密，果汁多；可溶性固形物含量15.5%，可滴定酸含量0.40%，果肉硬度12.5kg/cm²。树体矮小，树势强健，树冠紧凑，以短果枝结果为主，丰产性好。果实发育期170d左右。可在富士栽培区栽植。

83. Fudao Spur

Origin   Fudao Spur is a good bud mutation of Fuji apple varieties, It was introduced into China from Japan in 1984, and was examined and approved by Liaoning Crop Cultivar Registration Committee in 1993.

Main Characters   The fruit shape is near globose. Its average fruit weight is about 250g. The skin of fruit is red color, good skin finish and small dots. The flesh is in yellow white color and in good brittleness. The soluble solid substance content is 15.5%, the total titratable acid 0.40%, the firmness 12.5kg/cm². The tree of Fudao Spur is small and vigor, the canopy is compact with short branch, it is also very productive. The fruit developing period is about 170 days. Suitable for the regions where Fuji is planted.

84. 嘎拉

来源　新西兰以Kidd's Orange Red × 金冠杂交育成的中熟品种，1939年选出，1960年发表，1979年从日本青森县引入我国。目前嘎拉的浓红型芽变品种在世界各地均广泛应用，在我国亦应用面积较大。

主要性状　果实短圆锥形，平均单果重144.9g，底色淡黄，阳面着鲜红色条纹，果面光洁，果点中大、褐色；果肉淡黄色，肉质松脆，汁液多，风味甜，有香气；可溶性固形物含量13.4%，果肉硬度7.7kg/cm²。果实发育期135d，采后较耐贮藏。幼树树势强，进入结果期树势中庸，易成花，坐果率高，丰产性强。接种鉴定表现中抗苹果枝干轮纹病，中感苹果果实轮纹病，感苹果腐烂病和斑点落叶病。

84. Gala

Origin　Gala is a mid-season cultivar selected from hybrids of Kidd's Orange Red × Golden Delicious in New Zealand in 1939. It was released in 1960 and introduced into China from Aomori Prefecture, Japan, in 1979. Now, Gala and its red colored mutant cultivars have been widely used all over the world, including China.

Main Characters　The fruit shape of Gala is truncate conical and the fruit size is 144.9g. The ground color is light yellow and the sun-side is covered with striped Turkey red. The fruit skin is lubricity. The cuticular dot is large in size and brown in color. The flesh is ligh yellow in color, crisp in texture, much in juice, sweet in flavor and weak in aroma. The soluble solid content is 13.4%. The firmness without skin is 7.7kg/cm². The tree vigor is strong at yong stage and becomes moderate when fruiting. The flowering ability is strong. The fruit setting ratio and productiveness are high. The number of days of fruit growth is about 135 and the storage ability is fairly strong after harvest. Gala is moderately resistant to *Botryosphaeria* canker, moderately susceptible to apple fruit ring rot and susceptible to apple *Valsa* canker and *Alternaria* leaf blotch evaluated by inoculation with the pathogens.

85. 工藤

来源 日本从富士中选出的芽变品种。目前在北京昌平等苹果产区应用面积较大。

主要性状 果实近圆形；平均单果重260g；底色黄绿，着红色条纹；果肉淡黄色，肉质细脆，汁液多，风味甜；可溶性固形物含量15.0%。果实发育期180d左右，采后耐贮藏性强。树势中庸，较易成花，坐果率高，丰产性好。

85. Miya

Origin Miya is a mutant selected from Fuji and now has been widely used at some producing areas like Changping District, Beijing.

Main Characters The fruit shape of Miya is near globose and the fruit saze is 260g. The ground color is yellowish-green and is covered with striped red. The flesh is light yellow in color, fine in coarseness, crisp in texture, much in juice and sweet in flavor. The soluble solid content is 15.0%. The tree vigor is moderate. The floral bud formation is fairly easy. The fruit setting ratio and productivity are high. The number of days of fruit growth is about 180 and the storage ability is strong.

86. 宫崎短枝富士

来源　日本宫崎县从着色系富士中选育的短枝型芽变品种，1974年育成，1979年由中国侨联妇女代表团引入我国。目前在我国苹果产区应用面积较大，为主栽品种之一。

主要性状　果实近圆形；平均单果重240g；底色黄绿，着红色晕；果面光滑，果点小；果肉淡黄色，肉质细脆，汁液多，风味甜；可溶性固形物含量14.4%，可滴定酸含量0.38%，果肉硬度11.40kg/cm^2；果实发育期180d左右，果实采后耐贮藏性强。树势中庸，短枝型性状明显，较普通型富士品种易成花，坐果率高，丰产。田间表现抗早期落叶病，感苹果腐烂病、枝干轮纹病和果实轮纹病；接种鉴定均表现感枝干轮纹病，中抗斑点落叶病。

86. Miyazaki Spur Fuji

Origin　Miyazaki Spur Fuji is a spur type mutant selected from Red Fuji in Miyazaki Prefecture, Japan, in 1974. It was introduced into China by Chinese Women Federation Delegation in 1979 and now has been widely commercially used in China.

Main Characters　The fruit shape of Miyazaki Spur Fuji is near globose and the fruit size is about 240 g. The ground color is yellowish-green and is covered with splashed red. The fruit skin is lubricity. The cuticular dot is small is size. The flesh is light yellow in color, fine in coarseness, crisp in texture, much juicy and sweet in flavor. The firmness without skin is 11.40kg/cm^2. The soluble solid content and total titratable acid content are 14.4% and 0.38%, respectively. The number of days of fruit growth is about 180. The storage ability is strong. The tree vigor is moderate. Miyazaki Spur Fuji is apparent spur type. The floral bud formation is easier comparing to none spur type Fuji. The fruit setting ratio and productivity are high. The field resistance to leaf blotchs is high but is susceptible to apple *Valsa* canker, *Botryosphaeria* canker and fruit ring rot. Miyazaki Spur Fuji is susceptible to *Botryosphaeria* canker and moderately resistant to apple *Alternaria* leaf blotch evaluated by inoculation with the pathogens.

87. 国光

来源 产地及起源不详。多认为原产于美国弗吉尼亚州，起源于自然实生。

主要性状 果实扁圆形；平均单果重130g；底色黄绿，有红至暗红晕和宽短断续红条纹；果面光滑，果点多，中等大，明显；果肉乳白色，风味酸甜或甜酸；可溶性固形物含量13.0%，可滴定酸含量0.77%，果肉硬度9.6kg/cm²；肉质松脆，汁多味浓，品质中上，为鲜食品种。果实发育期170d左右。结果较晚，丰产性好。耐寒、耐旱、抗风，不易落果，对苹果树腐烂病抵抗力弱，易感锈果病。

87. Ralls

Origin   Ralls as a seedling with unclear parents, maybe originated in America Virginia.

Main Characters   The fruit shape is oblate. Its average fruit weight is 130g. The fruit ground color is yellowish-green with red to dark red and wide short intermittent stripe. The fruit surface is smooth with many middle-size obvious dots. The flesh is in milk white color. It tastes sour and sweet or sweet and sour. The soluble solid substance content is 13.0%, the total titratable acid is 0.77%, the firmness is 9.6kg/cm². It is late bearing, productive, crispy, juicy, tasteful and qualitative. It is suitable for fresh eating .The fruit developing period is about 170 days. It is tolerant to cold weather, drought and wind. The fruit is not easy to drop. It is weak in resistance of the apple tree canker and susceptible to rust disease.

88. 黑系乔纳金

来源　日本从乔纳金选育的芽变品种，21世纪初引入我国，2004年通过辽宁省品种审定。目前在我国辽宁西部有应用。

主要性状　果实圆锥形，平均单果重350g；底色黄绿，着浓红条纹，果点小；果肉乳黄色，肉质松脆，汁液多，风味酸甜，有香气；可溶性固形物含量15%。果实发育期165d，耐贮藏性较强。树势、结果习性及抗病性同乔纳金。

88. Excel Jonagold

Origin　Excel Jonagold is a mutant selected from Jonagold in Japan and was registered in Liaoning province, 2004. It was introduced into China at the beginning of this century and now has been used in the western Liaoning.

Main Characters　The fruit shape of Excel Jonagold is conical and the fruit size is 350g. The ground color is yellowish-green and is covered with striped strong red. The flesh is cream-yellow in color, crisp in texture, much in juice and sour-sweet. The soluble solid content is 15%. The fruit developing period is about 165 days. Some other characteristics, such as tree vigor, bearing habit and disease-resistance are similar to Jonagold.

89. 轰系津轻

来源 日本选育的中晚熟品种，来源于津轻的着色芽变。1974年命名发表，1979年从日本引入我国。

主要性状 果实近圆形，平均单果重200g；底色黄绿色，盖色鲜红或深红色，并有断续深红色条纹；果面光滑无锈，富有光泽，果点小；果肉黄白色，风味酸甜；可溶性固形物含量12.3%，可滴定酸含量0.24%，果肉硬度13.7kg/cm²；风味好，适于鲜食。果实发育期115d左右。丰产性强。适应性强，易栽培，好管理，但有采前轻微落果现象。

89. Hongxi Jinqing

Origin Hongxi Jinqing is a middle late apple variety selected from the colcured bud mutation of Jinqing in Japan, which was released in 1974 and introduced into China in 1979.

Main Characters The fruit shape is near globose. Its average fruit weight is about 200g. The ground color of skin is yellow green, and the cover color is turkeyt red or purplish red with dark red stripes. The fruit surface is smooth with no rust and small dots. The flesh is yellow white with sour and sweet flavor. The content of soluble solid is 12.3%, and the content of total titratable acid is 0.24%. The flesh hardness is 13.7kg/cm². Hongxi Jinqing is very productive, with good flavor, and suitable for fresh fruit. The fruit maturation is about 115 days. The cultivar has high adaptability, can be cultivated and managed easily. But there is slight fruit drop before harvest.

90. 弘前富士

来源　日本青森县北郡板柳町富士果园中发现的易着色极早熟富士品种。

主要性状　果实近圆形，果形端正，果形指数0.83；果个大，平均单果重248g；果面呈条状鲜红色，果点圆形；果肉黄白色，汁多、松脆、酸甜适中，可溶性固形物含量16.2%，果肉硬度10.9～12.5kg/cm^2；品质佳，耐贮性同富士。果实发育期145d左右，9月上中旬成熟，成熟期比富士早35～40d，比红将军早10d。

90. Hirosaki Fuji

Origin　Hirosaki Fuji is a easily colored and　early ripening variety found from Fuji orchard in Aomori Prefecture of Japan.

Main Characters　The fruit shape is near globose. The fruit shape index is 0.83. It is large in size and the average fruit weight is about 248g. The skin of fruit is striped on bright red base with small dots. The flesh is in light yellow color, juicy, crunchy and in good balance of sugar and acids. The soluble solid substance content is 16.2%, the firmness between 10.9 and 12.5kg/cm^2. The fruits store very well as Fuji. The fruit developing period is about 145 days. The fruits ripen in early September, 35~40 days earlier than Fuji, 10 days earlier than Benin Shogun.

91. 红安卡

来源   辽宁省果树科学研究所2002年从日本引进的富士实生苹果品种，2006年通过辽宁省非主要农作物品种备案办公室备案。

主要性状   果实近圆形或圆锥形，单果重270g，大果重310g。果实全红，果面平滑，无锈，果点小、少，无果粉。果肉黄白色，味甜，可溶性固形物含量15.4%，可滴定酸含量0.24%，果肉硬度8.4kg/cm^2；肉质松脆、中粗，汁多，微香，品质上等，耐贮藏，为鲜食品种。果实发育期165d左右。抗寒性中等，抗轮纹病能力强于富士，适宜在红富士苹果栽培区内发展。

91. Beni Aika

Origin   Beni Aika was first introduced from Japan by Liaoning Research Institute of Pomology in 2002. It was selected from Fuji seedlings. It was examined and approved by Liaoning Crop Cultivar Registration Committee in 2006.

Main Characters   The fruit shape is oblate or conical. Its average fruit weight is 270 g. The biggest fruit weight is 310g. The skin of fruit is in full red color. The fruit surface is smooth without rust and powder. The dots are small and a little. The flesh is in light yellow color and the taste is sweet. The soluble solid substance content is 15.4%, the total titratable acid is 0.24%, the firmness is 8.4kg/cm^2. It has intense aroma. Beni Aika is crispy, medium-textured, juicy, aroma and qualitative. It is long storage. It is suitable for fresh eating. The fruit developing period is about 165 days. It is middle tolerant to cold weather and more resistant to branch ring rot disease than Fuji. It is suitable for development in Fuji cultivation area.

92. 红盖露

来源　西北农林科技大学从美国引育的早熟新品种，为皇家嘎拉芽变。2006年通过了陕西省果树品种审定委员会审定。

主要性状　果实圆锥形；平均单果重180g；盖色浓红，着色有条纹，果皮光滑，果点大；果肉黄白色，风味酸甜适度；可溶性固形物含量14.6%，可滴定酸含量0.22%，果肉硬度11.3kg/cm^2；肉质硬脆，汁液多，有香气，为鲜食品种。果实发育期120d左右。丰产性好。红盖露品种树体适应性强，耐瘠薄，对早期落叶病、白粉病有一定的抗性。

92. Gale Gala

Origin　Gale Gala is a new early-maturing apple cultivar introduced by Northwest A&F University from the United States. It is a bud sport of Royal Gala. Gale Gala was examined and approved by Shaanxi Fruit Cultivar Registration Committee in 2006.

Main Characters　The fruit shape is conical. Gale Gala produces cone-shaped fruits with average fruit weight of 180 g. Its fruit skin is in full red color with dark red stripes, good skin finish and large dots. The flesh is in yellowish white color and in good balance of sugar and acids. Gale Gala contains 14.6% of the total soluble solid substance content, 0.22% of the total titratable acid, and its firmness is 11.3kg/cm^2. Gale Gala is a desert apple cultivar with high-yielding, crispy, juicy, tasteful characteristics. The fruit developing period of Gale Gala is about 120 days. In addition, Gale Gala does well in poor soil, tolerant to powdery mildew, and brown scab diseases.

93. 红将军

来源 日本选育的中熟品种，是从早生富士选出的着色系芽变。红将军曾称红王将。

主要性状 果实近圆形，果形指数0.86；平均单果重307g；果实色泽鲜艳，全面浓红；果肉黄白色，肉质细脆、多汁，风味甜酸浓郁，可溶性固形物含量15.9%，果肉硬度9.6kg/cm²，品质优。耐贮性强，不易发绵，自然贮藏可到春节。9月中旬成熟，比普通富士早熟30d以上。

93. Benin Shogun

Origin Benin Shogun used to be named as Hongwang jiang, is a mid ripening apple variety selected from the coloured bud mutation of Yataka Fuji in Japan.

Main Characters The fruit shape is near globose. The average fruit weight is about 307g and fruit shape index is 0.86. The skin of fruit is in full bright red color. The flesh is in white yellow color and is delicate, crispy, juicy, in good balance of sugar and acids. The soluble solid substance content is 15.9% and the firmness is $9.6kg/cm^2$. The fruits store well. The fruits ripen in mid-September, which is about 30 days earlier than Fuji.

94. 红星

来源　原产美国，为元帅系的第二代浓红芽变品种。1921年发现于新泽西州，1935年引入我国。

主要性状　果实圆锥形，五棱凸起明显，果形指数1.0左右；平均单果重230g，最大果重400g左右；初着色期呈浓红条纹，充分着色时全面浓红，但仍可见明显断续紫红条纹；果面光滑，有光泽，蜡质厚，果粉较多，果点浅褐色或灰白色，果肩起伏不平；风味酸甜，有芳香，品质上等，可溶性固形物含量13.3%左右，可滴定酸含量0.25%，果肉硬度7.7kg/cm^2；果肉淡黄色，松脆，汁多。果实发育期140d左右。耐贮性稍强于元帅，生理落果严重，采前落果较元帅轻。抗性较强。

94. Starking Delicious

Origin　Starking Delicious, originated from United States, is a mutation variety of a second generation variety of Delicious strain. It was found in State of New Jersey in 1921 and introduced into China in 1935.

Main Characters　The fruit is conical with angels on the fruit top and an average weight of 230g. Full strong red over color with significant intermittent striped purplish red, good surface finish, glossy, small dots and thick waxiness. The flavor is in intensely aromatic, tart-sweet with a soluble solid substance content of 13.3% and a total titratable acid content of 0.25%. The flesh is light yellow, crispy and very juicy with a firmness of 7.7kg/cm^2. The fruit developing period is about 140 days. The storability is better than Delicious. High resistant to diseases and serious physiological fruit drop.

95. 红玉

来源　原产美国，1980年在纽约州Ulster县发现，据说是可口香的实生后代，1926年发表。

主要性状　果实扁圆形；平均单果重145g；底色黄绿，阳面浓红色，充分着色可全面浓红；果面平滑，有光泽；果点小，较明显；果肉黄白色，初采时酸味较大，贮藏月余甜酸适口，风味浓郁；可溶性固形物含量13.0%，可滴定酸含量0.94%；肉质致密而脆，汁多味浓，品质上等，为鲜食和加工兼用品种。果实发育期150d左右。结果较早，较丰产。风土适应性强，但抗病能力较差，易感斑点落叶病、苹果树腐烂病、白粉病，采前落果稍重。

95. Jonathan

Origin　Jonathan is Native America. It was found Ulster County of New York Prefecture in 1980. It is said to be Kekouxiang's seedlings. It was realeased in 1926.

Main Characters　The fruit shape is oblate. Its average fruit weight is 145 g. The fruit ground color is yellowish-green with full blush. The fruit surface is smooth and shining with small obvious dots. The flesh is in light yellowish white color. It tastes sour when initial picking, It is in good balance of sugar and acids after 1 month storage. The soluble solid substance content is 13.0%, the total titratable acid is 0.94%. Jonathan is early bearing and productive. The flesh is fine and crispy, juicy, tasteful and qualitative. It is suitable for fresh eating and processing. The fruit developing period is about 150 days. It is strong adaptability, but it is poor in the disease resistant ability. It is susceptibility to leaf spot disease, apple canker and powdery mildew. Preharvest drop is slightly heavier .

96. 皇家嘎拉

来源　新西兰从嘎拉品种中选出的着色类型芽变。

主要性状　果实短圆锥形，平均单果重136g；果实底色黄，果面着橙红或鲜红色晕；果肉淡黄色，肉质松脆，稍疏松，汁中多；可溶性固形物含量13.6%，酸甜味浓，芳香浓郁，品质上等。果实发育期120d左右。皇家嘎拉除着色比嘎拉鲜艳外，其他性状与嘎拉相似。

96. Royal Gala

Origin　Royal Gala is a mutation of Gala with better color selected by New Zealand.

Main Characters　The fruit shape is truncate conical. Its average fruit weight is 136g. The skin of fruit is orange red or bright red blush with yellow background color. The flesh is light yellow, crisp, juicy and little soft. The soluble solid content is 13.6%. The eating quality is good with rich sour sweet flavor and aromatic. The fruit developing period is about 120 days. It is almost no difference with Gala except the better skin color.

97. 黄冠

来源　西北农林科技大学从法国引育的中熟新品种，亲本为Golden Delicious × Pilot。2013年通过陕西省果树品种审定委员会审定。

主要性状　果实圆锥形；平均单果重310g；果实黄色，阳面着红晕；果面光滑，果点稀小；果肉黄白色，风味酸甜适度；可溶性固形物含量14.2%，可滴定酸含量0.27%，果肉硬度6.8kg/cm²；汁液多，品质优，为鲜食品种。果实发育期160d左右。丰产性好。适应性强，对早期落叶病和白粉病有一定的抗性。

97. Yellow Delicious

Origin　Yellow Delicious is a new middle-maturing apple cultivar introduced by Northwest A&F University from France. Its parentage is Golden Delicious × Pilot. Yellow Delicious was examined and approved by Shaanxi Fruit Cultivar Registration Committee in 2013.

Main Characters　The fruit shape is conical. Yellow Delicious produces cylinder-shaped fruits with average fruit weight of 310g. Its fruit skin is in yellow color with red stripes on sunny side, good skin finish, and small dots. The flesh is yellowish white and in good balance of sugar and acids. Yellow Delicious contains 14.2% of the total soluble solid substance content, 0.27% of the total titratable acid, and its firmness is 6.8kg/cm². Yellow Delicious is a desert apple cultivar with high yields and juicy, and tasteful characteristics. The fruit developing period of Yellow Delicious is about 160 days. Yellow Delicious does well in various conditions, tolerant to powdery mildew and brown scab diseases.

98. 黄太平

来源 原产俄国西伯利亚，由耶弗列莫夫从大苹果自然授粉种子实生树中选出，一说由海棠种子实生而来。1909年传至哈尔滨，随后逐渐扩散到东北各地广泛栽培。20世纪50年代又传至内蒙古东部、河北北部、山东北部及新疆等地，是中国北方寒冷地区分布最广、栽培面积最大的小苹果品种。

主要性状 果实扁圆形，平均单果重32.9g。底色黄，阳面有鲜红晕，颇美观。果面光滑，有浅的棱起，无蜡质，富光泽；果点小，稀疏不明显。果肉浅黄色，风味甜酸，微涩；可溶性固形物含量13.5%，可滴定酸含量0.55%，不耐贮藏。肉质致密而脆；汁液多，品质中等，是既可鲜食又可加工的小苹果优良品种。果实发育期95d左右。早果丰产。黄太平植株抗寒力很强，不抗黑星病和早期落叶病，盛果期后易感染苹果树腐烂病，但刮治后恢复很快。

98. Huangtaiping

Origin Huangtaiping is selected from big apple seedling in Siberian of Russia by Jerzy F Leo Mo J. It is another said it was from Malus spectabilis seedling. It was spread to Haerbin in 1909, then it was widely planting in Northeast area gradually. It was spread to east of Inner Mongolia, north of Hebei and Xinjiang in the 1950s. It was the biggest planting area in distributing of cold area in North China.

Main Characters The fruit shape is oblate. Its average fruit weight is 32.9g. The fruit ground color is yellow. The skin of fruit is red blush, it is very slinky. The fruit surface is smooth with shallow ridge. It is no waxiness. It is rich in shinning. Small dots and sparsely. The flesh is in light yellow color and in taste of sugar and acids. It is a little acerbity. The soluble solid substance content is 13.5%, the total titratable acid is 0.55%. It is weak in storage. Huangtaiping is early bearing and productive, the flesh is fine and crispy, juicy, middle qualitative. It is suitable for fresh eating and processing. The fruit developing period is about 95 days. It is strong tolerant to cold weather. It is not resistant to scab and early defoliation disease. It is easy infectious apple canker after high productive period, but it recovery soon after scratch control.

99. 金矮生

来源　原产美国，1960年G.奥维尔在美国华盛顿州发现的金冠品种的短枝型芽变。1974年中国农业科学院从波兰和阿尔巴尼亚分别引入。

主要性状　果实圆锥形；平均单果重200g；果面光滑，洁净，果点中小。可溶性固形物含量15%，果肉硬度6.2kg/cm^2；肉质松脆，汁液中多，风味同普通金冠。果实发育期150d左右。丰产性好。金矮生比金冠树矮小，结果更早，适应能力也较金冠更强些。

99. Gold Spur

Origin　Gold Spur is a spur bud mutation of Golden Delicious, and is originated fund from the United States in 1960. It was introduced into China from Poland and Albania in 1974 by Chinese Academy of Agricultural Sciences.

Main Characters　The fruit shape is conical. Its average fruit weight is about 200g. The skin of fruit is finish with small dots. The flesh is in light yellow color and crispy. The soluble solid substance content is 15%, the firmness 6.2kg/cm^2. Gold Spur is very productive, crispy, juicy, tasteful similar to Golden Delicious. The fruit developing period is about 150 days. Compare to Golden Delicious, the trees of Gold Spur is shorter, earlier fruit, and better adaptive capacity.

100. 金冠

来源 美国以自然实生选育的中晚熟品种，1916年发表，1930年引入我国。

主要性状 果实圆锥形，平均单果重184g；果面金黄色，阳面稍具红晕；果面稍粗糙，果点大；果肉黄白色，风味酸甜适度；可溶性固形物含量14.6%，可滴定酸含量0.52%，果肉硬度6.4kg/cm²；肉质细脆，具浓郁芳香，适于鲜食。果实发育期150d左右。丰产性好。树体适应性强，栽培范围广，但果实易生果锈。

100. Golden Delicious

Origin   Golden Delicious is a middle late variety selected from natural seedlings in America, which was released in 1916 and introduced into China in 1930.

Main Characters   The fruit is conical. Its average fruit weight is about 184g. The fruit surface is in golden yellow with slightly red on the sunward side, a little rough and with big dots. The flesh is yellow white and with moderately sweet and sour flavor. The content of soluble solid is 14.6%, and the content of total titratable acid is 0.52%.The flesh hardness is 6.4kg/cm². Golden Delicious is very productive, crispy, with strong aroma, and suitable for fresh fruits. The fruit maturation is about 150 days. The cultivar with high adaptability can be widely cultivated, but sometimes there are fruit russeting on the fruit surface.

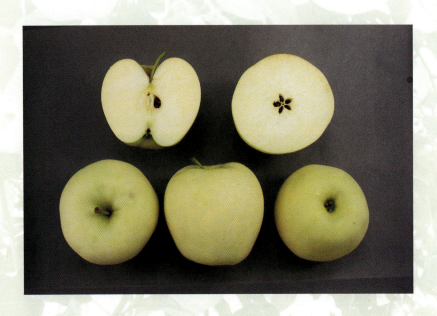

101. 津轻

来源   日本青森县苹果试验场以金冠 × 红玉杂交选育的中早熟品种，1930年杂交，1943年选出，1973年以青森2号发表，1975年以津轻品种登录，1979年从日本引入我国。目前津轻的浓红型芽变品种在日本应用较多，在我国各苹果产区亦有少量应用。

主要性状   果实圆形，平均单果重165g，底色黄绿，着红色条纹，果点中等大，白色或锈色；果肉乳白色，肉质较脆、汁液多，风味甜。可溶性固形物含量15.5%，果肉硬度6.83kg/cm^2。果实发育期110d，采后贮藏性较强。幼树树势强，结果后树势中庸，萌芽率中等，成枝力强，易成花，坐果率高，丰产性强，采前落果较重。田间表现抗苹果斑点落叶病和苹果果实轮纹病；接种鉴定表现中抗苹果果实轮纹病，感苹果枝干轮纹病和斑点落叶病。

101. Tsugaru

Origin   Tsugaru is a mid-season cultivar selected from hybrids of Golden Delicious × Jonathan at Aomori Prefectural Apple Experimental Station, Japan. The cross was made in 1930. This cultivar was selected in 1942, released in 1973 with the name Aomori No.2 and was registered in 1975 as Tsugaru. It was introduced into China in 1979. The red mutant cultivars of Tsugaru now have been commercially used widely in Japan and also in some apple producing areas of China.

Main Characters   The fruit shape of Tsugaru is globose and the fruit size is 165 g. The ground color is yellowish-green and is covered with striped red. The cuticular dot is medium in size and white or rusty in color. The flesh is cream-white in color, fairly crisp in texture, much in juice and sweet in flavor. The soluble solid content is 15.5%. The firmness without skin is 6.83kg/cm^2. The number of days of fruit growth is about 110 and the storage ability is fairly strong after harvest. The tree vigor is strong at young stage and becomes moderate when fruiting. The sprouting ratio is intermediate. The shooting ability and the floral bud formation ability are strong. The fruit setting ratio and the productivity are high. The field resistance is high to apple *Alternaria* leaf blotch and fruit ring rot, but Tsugaru is moderately resistant to apple fruit ring rot, susceptible to *Botryosphaeria* canker and *Alternaria* leaf blotch evaluated by inoculation with the pathogens.

102. 橘苹

来源　英国以Ribston Pippin的种子实生育成的中熟品种。1850年发表，20世纪20年代中期引入我国。

主要性状　果实扁圆形，平均单果重150g；盖色橙红色，并有浅红条纹；果面较粗糙，果点大；果肉黄白色，风味酸甜；可溶性固形物含量12.5%，可滴定酸含量0.37%，果肉硬度7.6kg/cm²；汁液中多，风味较浓，稍有香气，适于鲜食。果实发育期120d左右。丰产性强。适应性不强，不抗炭疽病。

102. Cox's Orange Pippin

Origin　Cox's Orange Pippin is a mid-maturation apple variety selected from Ribston Pippin seedlings in England, which was released in 1850 and brought into China in mid 1920s.

Main Characters　The fruit shape is oblate. Its average fruit weight is about 150 g. The cover color of skin is orange red with light red stripes. The fruit surface is rough with big dots. The flesh is yellow white with sweet and sour flavor. The content of soluble solid is 12.5%, and the content of total titratable acid is 0.37%.The flesh hardness is 7.6kg/cm². Cox's Orange Pippin is very productive, moderately juicy with good flavor, slightly aromatic, and suitable for fresh fruit. The fruit maturation is about 120 days. The cultivar has poor adaptability and sensitive to anthracnose.

103. 凯蜜欧

来源　西北农林科技大学从美国引进的中晚熟新品种，从自然实生苗中发现，亲本不详。2009年通过陕西省果树品种审定委员会审定。

主要性状　果实圆锥形；平均单果重300g；盖色红色，着色有条纹；果面光滑，果点小；果肉黄白色，风味酸甜适度；可溶性固形物含量14.9%，可滴定酸含量0.40%，果肉硬度9.5kg/cm²；肉质脆，有香气，品质优，维生素C含量高，为鲜食品种。果实发育期170d左右。丰产性好。树体适应性较强，对早期落叶病、白粉病和叶螨具有一定的抗性。

103. Cameo

Origin　Cameo is a new apple cultivar introduced by Northwest A&F University from the United States, maturing in early October. It originated as a seedling with unclear parents. Cameo was examined and approved by Shaanxi Fruit Cultivar Registration Committee in 2009.

Main Characters　The fruit shape is conical. Cameo produces cone-shaped fruits with average fruit weight of 300g. Its fruit skin is in full red color with dark red stripes, good skin finish and small dots. The flesh is in yellowish white color and in good balance of sugar and acids. Cameo contains 14.9% of the total soluble solid substance content, 0.40% of the total titratable acid, and relatively high vitamin C contents. Its firmness is 9.5kg/cm². Cameo is a desert apple cultivar with high yields, developed flavor, and crispy, juicy, and tasteful characteristics. The fruit developing period of Cameo is about 170 days. In addition, Cameo can be acclimated to various conditions. It is tolerant to powdery mildew, brown scab diseases, and spider mites.

104. 丽嘎拉

来源　新西兰从嘎拉中选出的着色系品种，原陕西省果树研究所于1995年引入我国试栽。

主要性状　果实近圆锥形，果形指数0.87；果个大，平均单果重220g，最大果重350g；较皇家嘎拉着色早，着色面大，片红，充分着色后全面浓红；果点明显，乳白色，果皮光滑，果粉多，有光泽，无锈；风味浓，可溶性固形物含量13.6%，果肉硬度7.2kg/cm^2；果肉淡黄色，肉质硬脆，汁液中多；耐贮性优于皇家嘎拉。在陕西渭北地区8月上中旬成熟。抗病性较强。

104. Regal Gala

Origin　Regal Gala is an early season and colored superior apple variety, originated from Gala in New Zealand. It was introduced into China by Research Institute of Pomology in Shaanxi Province in 1995.

Main Characters　The fruit is conical with a fruit shape index 0.87, large in size with an average weight of 220g. Complete over strong red with good surface finish, white dots and thick fruit powder. The flavor is intense with a soluble solid substance content of 13.6%. The flesh is light yellow, fine, juicy and crisp with a firmness of 7.2kg/cm^2, better storability than Rogal Gala. High resistance to diseases.

105. 凉香

来源　辽宁省果树科学研究所于1999年从日本引进的富士实生晚中熟苹果品种。2006年通过辽宁省非主要农作物品种备案办公室备案。

主要性状　果实近圆形，平均单果重325g，果个整齐，果实底色黄绿，全面着鲜红色。果面光滑，果点小；果肉淡黄色，风味酸甜，芳香浓郁。可溶性固形物含量15.4%，可滴定酸含量0.22%，果肉硬度7.5kg/cm²；肉质松脆，汁液多，味浓，品质优，为鲜食品种。果实发育期145d左右。丰产性好。抗性与富士相当，适宜在富士栽培区发展。

105. Ryoka no Kisetsu

Origin　Ryoka no Kisetsu was first introduced from Japan by Liaoning Research Institute of Pomology in 1999. It is a new late-middle ripening apple variety selected from Fuji seedlings. It was examined and approved by Liaoning Crop Cultivar Registration Committee in 2006.

Main Characters　The fruit shape is oblate. Its average fruit weight is 325 g. The fruit is regular. The fruit ground color is yellowish-green. The skin of fruit is in full red color. The fruit surface is smooth with small dots. The flesh is in light yellow color and the taste is sour and sweet. It has intense aroma. The soluble solid substance content is 15.4%, the total titratable acid is 0.22%, the firmness is 7.5kg/cm². Ryoka no kisetsu is very productive, crispy, juicy, tasteful and qualitative. It is suitable for fresh eating. The fruit developing period is about 145 days. Its resistibility is equal to Fuji. It is suitable for development in Fuji cultivation area.

106. 陆奥

来源 日本青森县苹果试验场以金冠 × 印度杂交育成的苹果晚熟品种。染色体倍性为三倍体。1930年杂交，1949年发表，20世纪60年代引入我国，目前在日本和欧洲生产上有应用，我国栽培少。

主要性状 果实近圆形；果顶偶有棱起，单果重260～310g；果点中多；不套袋果实果面绿色，贮藏后为金黄色；套袋后着鲜红色晕；果肉乳黄色，肉质较粗，松脆，汁液多，风味酸甜；可溶性固形物含量13.2%，可滴定酸含量0.58%，果肉硬度9.3kg/cm^2；耐贮性较强。陆奥是综合性状优良的鲜食烹饪兼用型苹果品种。花粉量少，果实发育期165d左右。幼树树势强，枝条粗壮，盛果期树势中庸，结果早，易成花，连续结果能力强，丰产。田间表现较抗苹果轮纹病和苹果腐烂病；接种鉴定表现中抗苹果枝干轮纹病，高感苹果腐烂病、苹果斑点落叶病。

106. Crispin

Origin Crispin is a late season triploid apple cultivar selected from hybrids of Golden Delicious × Indo at Aomori Prefectural Apple Experimental Station, Japan. The cross was made in 1930 and this cultivar was released in 1949. Crispin was introduced into China in 1960s and now has been used in Japan and some European countries but less used in China.

Main Characters The fruit shape of Crispin is near globose, the fruit size ranges from 260 to 310g. The fruit cuticular dot is dense and is medium in size. The fruit ground color appears yellow green without bagging and turns to golden yellow after storage, but is covered with splashed Turkey red in bagged fruit. The flesh is light yellow in color, medium in coarseness, crisp in texture. The flavor of flesh is sour-sweet with a good balance of sugars and acids. The firmness without skin is 9.3kg/cm^2. The soluble solid content and total titratable acid content are 13.2% and 0.58% respectively. It is elite for both fresh market and cooking. The tree vigor is strong at young stage and becomes medium when fruiting. The everbearing ability is strong but the pollen is almost sterile. Crispin is early bearing and highly productive. The number of days of fruit growth is about 165 and the storability of the fruit is fairly good. The field resistance to apple ring rot and apple *Valsa* canker is relatively high, but Crispin is moderately resistant to *Botryosphaeria* canker and is highly susceptible to *Valsa* canker and apple *Alternaria* leaf blotch evaluated by inoculation with the pathogen.

107. 美八

来源   美国纽约州农业试验站从嘎拉的杂交后代中选出来的优系，代号NY543。中国农业科学院郑州果树研究所于1984年从美国引入。

主要性状   果形近圆形，平均单果重180g；果面光洁无锈，底色乳黄，着鲜红色霞；果肉黄白，肉质细脆，多汁，可溶性固形物含量12.4%，风味酸甜适口，香味浓，品质上等。果实发育期110d左右，比嘎拉早成熟10d左右。

107. Meiba

Origin   Meiba is one of elite strains from the progenies of Gala by The New York State Agricultural Experiment Station. The original code is NY543. It was introduced to China by Zhengzhou Fruit Research Institute, Chinese Academy of Agricultural Sciences in 1984.

Main Characters   The fruit shape is near globose. Its average fruit weight is about 180 g. The surface of fruit is smooth, bright and clean, covered with bright red blush on the cream yellow background color. The flesh is yellow white color, fine, crisp and juicy. The soluble solid content is 12.4%. The eating quality is good with suitable sour sweet flavor and strong aromatic. The fruit developing period is about 110 days. The mature period is earlier ten days than Gala.

108. 美味

来源　西北农林科技大学从加拿大引育的中熟新品种，从自然实生苗中发现，亲本不详。2013年通过陕西省果树品种审定委员会审定。

主要性状　美味果实圆柱形；平均单果重270g；盖色红色，着色有条纹；果面光滑，果点小；果肉淡黄色，风味甜；可溶性固形物含量13.3%，可滴定酸含量0.14%，果肉硬度8.5kg/cm²；肉质细脆，汁液多，品质优，为鲜食品种。果实发育期155d左右。丰产性好。耐瘠薄，易管理，抗寒性强，对早期落叶病有一定的抗性。

108. Ambrosia

Origin　Ambrosia is a new middle-maturing apple cultivar introduced by Northwest A&F University from Canada. It originated as a seedling with unclear parents. Ambrosia was examined and approved by Shaanxi Fruit Cultivar Registration Committee in 2013.

Main Characters　The fruit shape is oblong. Ambrosia produces cylinder-shaped fruits with average fruit weight of 270 g. Its fruit skin is in full red color with dark red stripes, good skin finish and small dots. The flesh is sweet and pale yellow. Ambrosia contains 13.3% of the total soluble solid substance content, 0.14% of the total titratable acid, and its firmness is 8.5kg/cm². Ambrosia is a desert apple cultivar with high yields and crispy, juicy, and tasteful characteristics. The fruit developing period of Ambrosia is about 155 days. Ambrosia does well in poor soil, tolerant to cold-hardiness, and brown scab disease.

109. 蜜脆

来源　西北农林科技大学从美国引育的中熟新品种，亲本为Macoun × Honeygold。2006年通过陕西省果树品种审定委员会审定。

主要性状　果实圆锥形；平均单果重310g；盖色鲜红色，着色有条纹；果面光滑，果点小；果肉乳白色，风味微酸，有蜂蜜味；可溶性固形物含量15.0%，可滴定酸含量0.41%，果肉硬度9.20kg/cm²；肉质极脆，汁液特多，香气浓郁，品质优，为鲜食品种。果实发育期140d左右。丰产性好。树体抗寒性强，抗早期落叶病。

109. Honey Crisp

Origin　Honey Crisp is a new middle-maturing apple cultivar introduced by Northwest A&F University from the United States. Its parentage is Macoun × Honeygold. It was examined and approved by Shaanxi Fruit Cultivar Registration Committee in 2006.

Main Characters　The fruit shape is conical. Honey Crisp produces cone-shaped fruits with average fruit weight of 310g. Its fruit skin is in full red color with dark red stripes, good skin finish and small dots. The flesh is in cream color and with honey-like acid taste. Honey Crisp contains 15.0% of the total soluble solid substance content, 0.41% of the total titratable acid, and its firmness is 9.20kg/cm². Honey Crisp is a desert apple cultivar with productive, crispy, juicy, tasteful and other highly qualitative characteristics. The fruit developing period of Honey Crisp is about 140 days. In addition, Honey Crisp is tolerant to cold-hardness and brown scab disease.

110. 摩里士

来源　原产美国，由新泽西州农业试验场于1948年以（金冠×Edgewood）×（红花皮×克露丝）杂交选育而成的中早熟品种，1966年发表。我国先后由日本和美国引入试栽。

主要性状　果实短圆锥形，五棱突起明显；大果型，平均单果重300g，最大果重500g；底色黄绿，表皮着红霞或细红条纹，充分着色为全面浓红；果面光滑，有光泽，果粉中等，蜡质较厚；微香，酸甜适度，可溶性固形物含量13%～14%，可滴定酸含量0.21%；果肉乳黄色，肉质中粗，较松脆，汁液中多。果实发育期130d左右。不耐贮藏，室温下可存放1个月左右。连续结果能力较强，较丰产，抗病性较强。

110. Mollie's Delicious

Origin　Mollie's Delicious, native to United States, is a new mid-early season apple variety, originated from the cross (Golden Delicious × Edgewood) × NJ4 (Gravensteiner Red × Close) made by New Jersey Agricultural Experiment Station in 1948 and released in 1966. It was successively introduced into China from Japan and America .

Main Characters　The fruit is Truncate conical with angels on the fruit top, large in size with an average weight of 300 g. Yellow green ground color with striped red overcolor, good surface finish, medium fruit powder and thick waxiness. The flavour is aromatic, in good balance of sugar and acid with a soluble solid substance content of 13%~14% and a total titratable acid content of 0.21%. The flesh is cream yellow, mid-coarse, crispy and juicy. The fruit development period is about 130 days. High and stable yields, resistant to virus diseases and short storage for 1 month at room temperature.

111. 南方脆

　　来源　由新西兰以嘎拉 × 华丽杂交而成，原代号为GS330，原陕西省果树研究所于1995年引入。

　　主要性状　果实近圆锥形，果形指数0.85；果个中等，平均单果重160～180g，较嘎拉略大；果实全红，着色鲜艳，上色较嘎拉早；风味浓，可溶性固形物含量13.03%，果肉硬度8.3kg/cm²，耐贮性优于嘎拉。在陕西渭北地区8月中下旬成熟，与皇家嘎拉相近。抗病性较强。

111. Southern Snap

Origin　Southern Snap is an early-mid season variety, originated from the cross Gala × Splendour in New Zealand. It was introduced into China by Research Institute of Pomology, Shaanxi Province in 1995.

Main Characters　The fruit is conical with a fruit shape index of 0.85, medium in size with an average weight of 160~180g. Full red over color with earlier coloration than Gala. The flavor is intense with a soluble solid substance content of 13.03% and a firmness of 8.3kg/cm². High resistance to diseases, better storability than Gala.

第二部分　中国苹果栽培品种

112. 皮诺娃

来源 西北农林科技大学从德国引育的中熟新品种，亲本为Clivia（Oldenburg × Cox Orange）× Golden Delicious。

主要性状 果实圆形；平均单果重230g；盖色红色，着色有条纹；果面光滑，果点稀小；果肉黄白色，风味酸甜适度；可溶性固形物含量13.8%，可滴定酸含量0.40%，果肉硬度8.3kg/cm²；肉质细脆，汁液多，品质优，为鲜食品种。果实发育期160d左右。丰产性好。树体抗寒性强，对早期落叶病和白粉病有一定的抗性。

112. Pinova

Origin Pinova is a new middle-maturing apple cultivar introduced by Northwest A&F University from German. Its parentage is Clivia (Oldenburg × Cox Orange) × Golden Delicious.

Main Characters The fruit shape is globose. Pinova produces round-shaped fruits with average fruit weight of 230g. Its fruit skin is in full red color with dark red stripes, good skin finish and small dots. The flesh is in yellowish white color and in good balance of sugar and acids. Pinova contains 13.8% of the total soluble solid substance content, 0.40% of the total titratable acid, and its firmness is 8.3kg/cm². Pinova is a desert apple cultivar with high yields and crispy, juicy, and tasteful characteristics. The fruit developing period of Pinova is about 160 days. In addition, Pinova is tolerant to cold-hardness, powdery mildew, and brown scab diseases.

113. 千秋

来源　日本秋田县果树试验场以东光 × 富士杂交选育而成，1979年发表，1980年登记。1981年引入我国。

主要性状　果实圆形或长圆形，果形指数0.9；中果型，平均单果重160g，最大果重210g；底色绿黄，着浓红色彩霞和断续条纹，充分着色可达全红；果面光滑，有光泽，果点较小、稍稀，果锈少，果皮薄；有香气，风味酸甜，可溶性固形物含量13% ~ 14%，可滴定酸含量0.5%左右；果肉黄白色，肉质细、致密，汁液多。果实发育期145d左右，较耐贮藏。采前落果少，丰产、稳产。

113. Senshu

Origin　Senshu is a new apple variety, originated from the cross Toko × Fuji by Akita County Fruit Proving Ground in Japan. It was published in 1979, registered in 1980 and introduced into China in 1981.

Main Characters　The fruit is globose or oblong globose with a fruit shape index of 0.9, medium in size with an average weight of 160g. Yellow green ground color with striped red overcolor, good surface finish and small dots. The flavor is aromatic, sweet with a soluble solid substance content of 13%~14% and a total titratable acid content of 0.5%. The flesh is yellow white, fine and juicy. They are high and stable yields, 145 days fruit development period, storage life about 3 months.

114. 乔纳金

来源 美国纽约州Geneva农业试验站用金冠 × 红玉杂交育成的三倍体苹果品种，1943年杂交，1953年入选，1968年发表，1979年引入我国。目前乔纳金的浓红型芽变品种在欧美和日本有较多应用，我国亦有一定应用。

主要性状 果实圆锥形或短圆锥形，平均单果重280g；底色黄绿，着红色断续条纹；果面光滑，果点小；果肉乳黄色，肉质松脆，汁液多，风味酸甜，有香气；可溶性固形物含量15%。果实发育期165d，采后贮藏力中等，为优良的鲜食及烹饪兼用品种。树势强，萌芽率高，成枝力强，极易成花，有自花结实能力，丰产性强，连续结果能力强。田间表现对多种病害的抗病性均较强；接种鉴定表现中抗苹果果实轮纹病，感苹果枝干轮纹病，高感苹果腐烂病和斑点落叶病。

114. Jonagold

Origin Jonagold is a triploid apple cultivar selected from hybrids of Golden Delicious × Jonathan at Geneva Agricultural Experimental Station, New York, USA, in 1953. The cross was made in 1943 and this cultivar was released in 1968. It was introduced into China in 1979. Now, red mutation cultivars of Jonagold are widely used in Europe, the United States, Japan and are also commercially used in some area in China.

Main Characters The fruit shape of Jonagold is conical or truncate conical and the fruit size is 280 g. The ground color is yellowish-green and is covered with intermittent striped red. The fruit skin is lubricity. The cuticular dot is small. The flesh is cream-yellow in color, crisp in texture, much in juice, sour-sweet in flavor and rich in aroma. The soluble solid content is 15.5%. The tree vigor is strong. The sprouting ratio, shooting ability, and self-compatibility are high. The flower bud formation ability and the everbearing ability are strong. The productivity is high. The number of days of fruit growth is about 165 and the post harvest storage ability is intermediate. It is elite for both fresh market and cooking. The field resistance is relatively high to several diseases, but Jonagold is moderately resistant to apple fruit ring rot, susceptible to *Botryosphaeria* canker and highly susceptible to apple *Valsa* canker and *Alternaria* leaf blotch evaluated by inoculation with the pathogens.

115. 青森短枝富士

来源　日本青森县野村园艺农场育成的短枝富士品种。

主要性状　果实近圆形，平均单果重380g；果实颜色条状鲜红色，果点大而稀；果肉黄白色，肉质脆，致密，汁液多，甜酸适口，可溶性固形物含量16%，果肉硬度12.5kg/cm²。10月中下旬成熟。比普通富士品种抗寒性好。

115. Aomori Spur Fuji

Origin　Aomori Spur Fuji is a new spur type Fuji variety selected by Horticultural Farm of Nomura in Aomori of Japan.

Main Characters　The fruit shape is near globose. The average fruit weight is about 380 g. The skin of fruit is striped on red base with big dots. The dense flesh is in yellow-white color, juicy, crispy and in good balance of sugar and acids. The firmness is 12.5kg/cm², the soluble solid substance content 16%. The fruits ripen in mid to late October. It is tolerant to cold, hardier than Fuji.

116. 青香蕉

来源　美国育成的晚熟品种，起源不详，1849年已有记载，19世纪80年代传入我国。

主要性状　果实圆锥形或短圆锥形，平均单果重200g；底色绿或黄绿色，阳面偶有褐红晕；果面稍粗糙，果点密；果肉黄白色，风味甜酸；可溶性固形物含量13.0%，可滴定酸含量0.36%，果肉硬度10kg/cm^2；肉质较细，汁液较多，稍有芳香，可用于鲜食。果实发育天数为180d左右。树体抗寒性较差，但抗风力较强。

116. White Pearmain

Origin　White Pearmain is a late-maturing apple variety from America, which was record from1849 and brought into China in 1880s. Its origin is unknown.

Main Characters　The fruit shape is conical or truncate conical. Its average fruit weight is about 200 g. The ground color of skin is green or yellow green with brown red blush on the sunward side. The fruit surface is slightly rough with thick dots. The flesh is yellow-white with sweet and sour flavor. The content of soluble solid is 13.0%, and the content of total titratable acid is 0.36%. The flesh hardness is 10kg/cm^2. White Pearmain is juicy, slightly aromatic, and suitable for fresh fruit. The fruit maturation is about 180 days. It is tolerant to wind, but not to cold.

117. 清明

来源　日本秋田县平鹿町伊藤善明氏以金冠 × 富士杂交育成的苹果中晚熟品种，1995年登录，1994年由日本花甲协会会员菅井功氏引入河北省昌黎果树研究所。目前在我国生产应用不多。

主要性状　果实圆形至长圆形，单果重240～280g；果面光洁，果点锈色明显，果实着鲜红色条纹；果肉乳黄色，肉质脆，汁液多，风味甜；可溶性固形物含量15.8%，可滴定酸含量0.31%，果肉硬度6.4kg/cm^2；采后常温下可贮藏30d左右。果实发育期150d。树势中庸、早果、丰产性好，采前不落果。接种鉴定表现感斑点落叶病和苹果枝干轮纹病，高感苹果腐烂病。

117. Seimei

Origin　Seimei is a mid-late season apple cultivar selected from hybrids of Golden Delicious × Fuji by Yoshiaki Ito at Hiraka town, Akita Prefecture, Japan. Seimei was registered in 1995 and introduced into Changli Institute of Pomology, Hebei Academy of Agriculture and Forestry Sciences by Isao Sugai, a member of JSV, in 1994. Seimei is currently yet popularly used in China.

Main Characters　The fruit shape of Seimei is globose to oblong globose, the fruit size ranges from 240 to 280g. The fruit skin is lubricity covered with striped Turkey red. The cuticular dot is clear and is rusty in color. The flesh is cream-yellow in color, crisp in texture, much juicy and sweet in flavor. The firmness without skin is 6.4kg/cm^2. The soluble solid content and total titratable acid content are 15.8% and 0.31% respectively. The fruit can be stored at room temperature for about 30 days post harvest. The tree vigor is intermediate. Seimei is early bearing and highly productive. No pre-harvest fruit abscission. The number of days of fruit growth is about 150. Seimei is susceptible to apple *Alternaria* leaf blotch and *Botryosphaeria* canker, and is highly susceptible to *Valsa* canker evaluated by inoculation with the pathogens.

118. 瑞丹

来源 法国制汁专用苹果品种。

主要性状 单果重160g，果面黄绿带条红，果汁含量丰富，出汁率高达75%，可溶性固形物含量12.0%，原汁酸度0.36%，制汁品质佳。耐贮运，早实、丰产性强。枝干不抗轮纹病，果实成熟期为9月上旬。

118. Judaine

Origin   Judaine is an apple variety for juice processing in France.

Main Characters   The average fruit weight is about 160g. The fruit surface is striped on yellow-green base. The apple variety is juicy and the juice rate is about 75%. The soluble solid content is about 12.0%, the juice acidity 0.36%. The juice quality is well. The trunk and branch is susceptible to ring rot disease (*Botryosphaeria dothidea*). The trees are precocious and productive. The fruits ripen in early September.

119. 瑞林

来源 法国制汁专用苹果品种。

主要性状 单果重120g，果面绿色带条红，出汁率72%，可溶性固形物含量9.8%，原汁酸度0.30%，制汁优良，亦可鲜食。早实、丰产。9月上旬成熟。

119. Judeline

Origin　Judeline is an apple variety for juice processing in France.

Main Characters　The average fruit weight is about 120 g. The fruit surface is striped on green base. The juice rate is 72%, the soluble solid content is about 9.8%, the juice acidity 0.30%. The fruits are good for juice processing and also are suitable for fresh eating. The trees are precocious and productive. The fruits ripen in early September.

120. 珊夏

来源　日本和新西兰合作以嘎拉 × 茜杂交培育而成。

主要性状　果实圆形，平均单果重165g；果实底色黄绿，着淡红色；果肉乳白色，肉质细脆、汁液多；可溶性固形物含量13.2%，可滴定酸含量0.24%，风味酸甜，品质很好。果实发育期100～110d。该品种部分年份果面稍带有网状锈，而且叶片颜色稍淡，似缺肥状。

120. Sansa

Origin　Sansa is selected from collaborate breeding programme between Japan and New Zealand. Its parents is Gala × Akane.

Main Characters　The fruit shape is globose. Its average fruit weight is 165g. The skin of fruit is light red on the yellow green background color. The flesh is cream color, fine, crisp and juicy. The soluble solid content is 13.2%, the total titratable acid 0.24%. The eating quality is excellent with sour-sweet flavor. The fruit developing period is about 110~110 days. Fruits surface have a little bit reticular russet in some years, and the leaves is lighter than normal green color, just like fertilizer deficiency.

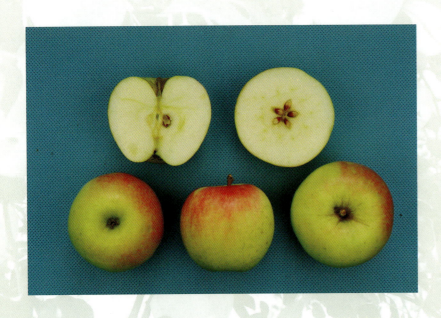

121. 盛放富1

来源 日本用放射线诱变富士选育的中熟品种。1981年引入我国。

主要性状 果实扁圆形，平均单果重300g；底色黄绿，盖色鲜红；果面平滑，果点大；果肉黄白，风味酸甜适度；可溶性固形物含量15.8%，果肉硬度6.8kg/cm²；肉质细脆，汁液多，适于鲜食。果实发育期138d左右。丰产性强。树体生长势强，适应性不强。

121. Shengfang Fu 1

Origin Shengfang Fu 1 is a mid-maturation apple variety selected from radiation mutation of Fuji in Japan, which was introduced into china in 1981.

Main Characters The fruit shape is oblate. Its average fruit weight is about 300g. The ground color of skin is yellow green, and the cover color is turkeyt red. The fruit surface is smooth with big dots. The flesh is yellow white with moderately sweet and sour flavor. The content of soluble solid is 15.8%, and the flesh hardness is 6.8kg/cm². Shengfang Fu 1 is productive, crisp, juicy, and suitable for fresh fruit. The fruit maturation is about 138 days. The cultivar has strong vigor and poor adaptability.

122. 首红

来源 原产美国，1967年在华盛顿州发现的新红星短枝型株变，为元帅系第四代芽变品种。1976年发表，1978年引入我国。

主要性状 果实长圆锥形，端正、高桩，五棱凸起明显；平均单果重280g，最大果重340g；底色绿黄，着全面浓红色，色泽艳丽，条纹不明显；果面有光泽，果粉少，蜡质较多，果点小；味甜，可溶性固形物含量14.2%，可滴定酸含量0.3%，有香气，无涩味；果肉乳白色，肉质细脆，汁液中多。果实发育期135d。植株紧凑，适于密植栽培。

122. Redchief Delicious

Origin   Redchief Delicious, originated from United States, is a spur type limb mutation of Starkrimson Delicious in 1967, recognizing as a fourth generation variety of Delicious strain. It was published in 1976 and introduced into China in 1978.

Main Characters   The fruit is narrow conical with angels on the fruit top and an average weight of 280 g. Green yellow ground color with prominent striped strong red over color, good surface finish, glossy, thick waxiness and small dots. The flavor is aromatic and no acerbity, sweet with a soluble solid substance content of 14.2% and a total titratable acid content of 0.3%. The flesh is cream, fine, crisp and medium juicy. The fruit developing period is about 135 days. Suitable for high density planting.

123. 松本锦

来源　日本以津轻×耐劳26号杂交而成的早熟品种，1993年引入我国，2000年通过山东省农作物品种审定委员会审定。

主要性状　果实圆至扁圆形，平均单果重280g；果面光洁艳丽，全面浓红；果肉黄白色，肉质细脆，汁液多；可溶性固形物含量13%～14%；风味酸甜适口，有香味，品质优良。果实发育期105d左右。该品种是一个红色、大果型早熟品种，田间表现易感苹果褐斑病。

123. Matsumotokin

Origin　Matsumotokin is an early ripening apple variety selected by Japan. Its parentage is Tsugaru × Nero 26. It was introduced to China in 1993 and was examined and approved by Shandong Crop Cultivar Registration Committee in 2000.

Main Characters　The fruit shape is globose or oblate. Its average fruit weight is 280g. The surface of fruit is smooth, bright and clean, covered full dark red color. The flesh is yellow white color, fine, crisp and juicy. The soluble solid content is 13%~14%. The eating quality is good with suitable sour sweet flavor and aromatic. The fruit developing period is about 105 days. It is a big size, red color and early mature variety. But it is susceptible to apple brown spot.

124. 太平洋玫瑰

来源　新西兰以嘎拉×华丽杂交育成的晚熟品种，原代号为GS2085，为GS系的代表性品种。我国20世纪90年代中期引入，2010年通过山东省品种审定。

主要性状　果实圆形，果形指数0.82；平均单果重230g；底色黄绿，表皮着鲜玫瑰红色；果皮薄，果面光亮，蜡质厚，外观艳丽；酸甜适口，有玫瑰香气，可溶性固形物含量15%，可滴定酸含量0.32%，果肉硬度8.8kg/cm²；果肉乳白色，肉质细脆，多汁；果实耐贮性好，自然条件下贮藏至翌年4月，硬度仍可达7.0kg/cm²。在陕西渭北地区，果实9月下旬成熟。坐果率高，无采前落果现象，丰产、稳产，但树体和果实易感褐斑病。

124. Pacific Rose

Origin　Pacific Rose is a late season variety, originated from the cross Gala × Splender in New Zealand. It was introduced into China in the 1990s and approved by Shandong Cultivar Registration Committee in 2010.

Main Characters　The fruit is globose with a fruit shape index of 0.82 and an average weight of 230g. Yellow green ground color with rose red over color, good surface finish, glossy and thick waxiness. The flavor is rose aromatic, in good balance of sugar and acid with a soluble solid substance content of 15.0% and a total titratable acid content of 0.32%. The flesh is cream, fine, juicy and crisp with a firmness of 8.8kg/cm². High yield stability and fruit setting rate with no fruit drop of pre-harvest. Susceptibility to brown spot.

125. 藤牧1号

来源　美国Purdue等3所大学由抗苹果黑星病和苹果白粉病育种项目中联合育成。20世纪70年代引入日本获得专利并进行试栽，在日本被命名为藤牧1号（MATO），又叫南部魁。1986年由日本引入我国。

主要性状　果实近圆形；平均单果重160g；果面光滑，底色黄绿，阳面有红晕；果肉黄白色，肉质细而松脆，汁液多，可溶性固形物含量12.2%，风味酸甜适口，有芳香，品质中上。果实发育期85～90d。

125. MATO

Origin　MATO is selected from the apple scab and powdery mildew resistance joint breeding project by American Purdue and other two universities and was introduced to Japan in 1970s. It was named MATO, also called Nanbukui in Japan, was introduced to China in 1986.

Main Characters　The fruit shape is near globose. Its average fruit weight is about 160g. The background color is yellow green and covered red blush on the sunny side. The flesh is yellow white color, fine, crispy and juicy. The soluble solid content is 12.2%. The eating quality is good with aromatic and good balance of sugar and acid. The fruit developing period is about 85~90 days.

126. 瓦里短枝

来源 原产美国，20世纪70年代末发现的首红芽变，为元帅系第五代短枝型品种，80年代中期引入我国。

主要性状 果实圆锥形，高桩，五棱明显；果个大，均匀整齐，平均单果重215g，最大果重350g；底色黄绿，全面着浓红霞彩或彩条红纹，果点稀少，鲜艳美观；风味甜酸适口，有香气，可溶性固形物含量12.8%，可滴定酸含量0.24%；果肉绿黄色，松脆，汁中多。果实发育期130d左右，贮藏性同新红星。树势中庸，属开张型短枝品种。

126. Vallee Spur Delicious

Origin Vallee Spur, originated from United States, is a spur type whole tree sport variety of Redchief Delicious in the 1970s, recognizing as a fifth generation variety of Delicious strain. It was introduced into China in the middle of 1980s.

Main Characters The fruit is conical with angels on the fruit top, large in size with an average weight of 215g. Yellow green ground color with full striped strong red over color and small dots. The flavor is aromatic, in good balance of sugar and acid with a soluble solid substance content of 12.8% and a total titratable acid content of 0.24%. The flesh is green yellow, loose, crispy and medium juicy. The fruit developing period is about 130 days. The storability is similar with Starkrimson Delicious.

127. 王林

来源 日本以金冠 × 印度杂交育成的晚熟品种，1952年命名。我国于1978年引入。

主要性状 果实椭圆形或卵圆形，平均单果重196.8g；底色为黄绿色或绿黄色，阳面略有红晕。果面较光滑，果点大明显；果肉乳白色，风味酸甜适度，有香气；可溶性固形物含量12.8%，可滴定酸含量0.27%，果肉硬度8.1kg/cm²；肉质细脆，汁液多，品质优，鲜食性状好。果实发育期155d左右。树体抗寒性较差，不抗斑点落叶病。

127. Orin

Origin Orin is a late-maturing apple variety bred by Golden Delicious × Indo in Japan, which was released in 1952 and brought into China in 1978.

Main Characters The fruit shape is ellipsoid or ovoid. Its average fruit weight is about 196.8g. The ground color of skin is yellow green or green yellow with slightly red on the sunward side. The fruit surface is smooth with big and obvious dots. The flesh is ivory with moderately sweet and sour flavor. The content of soluble solid is 12.8%, and the content of total titratable acid is 0.27%. The flesh hardness is 8.1kg/cm². Orin is crisp, juicy, tasteful, qualitative, and suitable for fresh fruit. The fruit maturation is about 155 days. The cultivar has poor tolerance to cold weather and Alternaria leaf spot.

128. 维斯特·贝拉

来源　美国新泽西农业试验站以77349 × 乔纳红杂交选育而成的早熟品种。

主要性状　果实扁圆或近圆形；平均单果重152g；果面底色绿黄，果面呈浓红色；果面光滑，果点大，果粉多；果肉乳白色，肉质松脆，汁液中多；可溶性固形物含量12.1%，可滴定酸含量0.72%，风味甜酸，品质中上。果实发育期70 ~ 75d。

128. Vista Bella

Origin　Vista Bella is an early ripening apple variety selected by New Jersey Agricultural Experiment Station. Its parentage is 77349 × Jonared.

Main Characters　The fruit shape is oblate or near globose. Its average fruit weight is 152g. The skin color is strong red and background color is green yellow. The fruit surface is smooth with big fruit dots. The flesh is cream color, crispy, and juicy. The soluble solid content is 12.1%, the total titratable acid 0.72%. The eating quality is medium to good with sweet-sour flavor and aromatic. The fruit developing period is about 70~75 days.

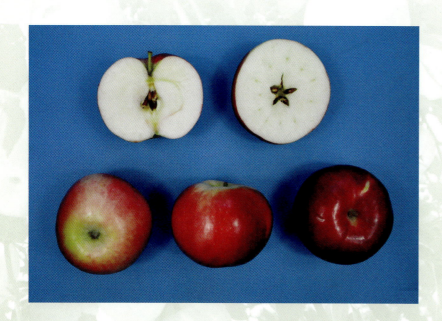

129. 倭锦

来源 原产美国，1871年引入我国。

主要性状 果实近圆形或短卵圆形，平均单果重235g；盖色鲜红色，并有浓红或暗红断续条纹；果面光滑，果点小；果肉白色，风味甜酸；可溶性固形物含量13.0%，可滴定酸含量0.35%，果肉硬度9.7kg/cm^2。果实发育期150d左右。丰产性强，花粉多，可作授粉品种。树势强，适应性强，较抗风，耐寒等。

129. Ben Davis

Origin Ben Davis is an apple variety from America, which was brought into China in 1871.

Main Characters The fruit shape is near globose or ovoid. Its average fruit weight is about 235g. The cover color of skin is turkeyt red with dark red stripes. The fruit surface is smooth with small dots. The flesh is white with sour and sweet flavor. The content of soluble solid is 13.0%, and the content of total titratable acid is 0.35%. The flesh hardness is 9.7kg/cm^2. The fruit maturation is about 150 days. Ben Davis is very productive with plenty of pollen, so suitable for pollination variety. The cultiver has strong vigor, high adaptability and strong resistance to wind and cold.

130. 舞佳

来源　舞佳（Polka）又名特珍（Trajan），是英国东茂林果树试验站于1976年以威赛克旭 × 金冠杂交选育出的柱型苹果品种，1986年发表。

主要性状　果实卵圆形至圆锥形，果形指数0.83；果个中大，平均单果重188g左右；果面底色绿黄，着红晕，红绿相间，着色不均匀；果面较平滑；果肉白色，肉质较细、脆、汁多，酸甜，风味浓，略有香气；可溶性固形物含量14.2%，果肉硬度11.5kg/cm^2，品质中上。除生食外可作为加工制汁品种。9月下旬成熟，耐藏性较差。

130. Polka

Origin　Polka, also known as Trajan, is a columnar apple variety selected by East Malling Research Station of Britain from a crossing between Mclntosh Wijcik and Golden Delicious made in 1976. It was published in 1986.

Main Characters　The fruit shape is ovoid to conical (a bit flat) and the fruit shape index is 0.83. The average fruit weight is 188g. The fruit skin is blushed on green-yellow base. The fruit surface is smooth and the flesh color is white. The fruit quality is above medium. It is crispy, juicy, acidic and aromatic. The flavor is good. The soluble solid content is 14.2% and the firmness 11.5kg/cm^2. The fruits can be used both for fresh eating and juice processing. The fruits ripen in late-September and can be stored for two months under the natural storage condition.

131. 舞乐

来源 舞乐（Bolero）又名塔斯坎（Tuscan），是英国东茂林果树试验站于1976年以威赛克旭×绿袖杂交选育出的柱型苹果品种，1986年发表。

主要性状 果实扁圆形，果形不甚端正，果形指数0.8；果个较大，平均单果重219g左右；果实底色绿，有轻微果锈，阳面有红色晕；果肉乳白色，肉质脆，果肉很易绵化，汁多、味酸甜，有香气；可溶性固形物含量11.0%，果肉硬度8.43kg/cm^2。除生食外可作为加工制汁品种。9月中旬成熟，耐藏性差。

131. Bolero

Origin Bolero, also known as Tuscan, is a columnar apple variety selected by East Malling Research Station of Britain from a crossing between McIntosh Wijcik and Greensleeves made in 1976. It was published in 1986.

Main Characters The fruit shape is oblate and the fruit shape index is 0.8.The fruit average weight is 219g. The fruit skin is in green color with little russets and blush. The fruit flesh color is milk-white. The fruits are crispy, juicy, acidic and rich in flavor. The soluble solid substance content is 11.0% and the firmness 8.43kg/cm^2. It is suitable both for fresh eating and juice processing. The fruits ripen in mid-September. The fruits do not store very well.

132. 舞美

来源 舞美（Maypole）又名玛宝（Maypole），是英国东茂林果树试验站于1976年以威赛克旭 × *M. baskatong* 杂交选育出的柱型苹果品种，1986年发表。

主要性状 果实圆形或圆锥形，果形指数0.91，果个小，平均单果重35.5g；成熟时果实底色绿黄，全面着橙红色，有红晕；果肉橘黄色，肉质较细，松软、不脆，汁液中等，风味酸，带涩味，可溶性固形物含量9.0%，品质下。不适于鲜食，可加工果汁、果酱。果实9月上旬成熟。属观赏树品种。

132. Maypole

Origin   Maypole, also known as Maypole, is a columnar apple variety selected by East Malling Fruit Experiment Station of British in 1976 and published in 1986. Its parentage is Mclntosh Wijcik × *M. baskatong* .

Main Characters   The fruit shape is globose or conical, the fruit is small in size, the fruit shape index is 0.91 and the average fruit weight 35.5g. The fruit skin is green yellow with full orange red and red blush. The flesh is soft, and tart. The soluble solid content is 9.0%. Therefore, the fruit is not suitable for fresh eating and can be used for juice and jam making. The fruits ripen in early September. It is an ornamental variety.

133. 舞姿

来源　舞姿（Waltz）又名特拉蒙（Telemon），是英国东茂林果树试验站于1976年以威赛克旭 × 金冠杂交选育出的柱型苹果品种，1986年发表。

主要性状　果实扁圆形，果形指数0.81，果个较大，平均单果重255g；果实底色绿黄，全面着深红彩色；果肉乳黄色，肉质细嫩，硬而脆，汁多、酸甜，略有香气，风味浓，可溶性固形物含量11.4%，果肉硬度9.98kg/cm²。品质中上。10月初成熟。较耐贮藏，在一般条件下可贮藏至翌年2月。

133. Waltz

Origin　Waltz, also known as Telemon, is a columnar apple variety selected by East Malling Research Station of Britain from a crossing between McLntosh Wijcik and Golden Delicious made in 1976. It was published in 1986.

Main Characters　The fruit shape is oblate and the fruit shape size is 0.81. The average fruit weight is 255g. The fruit skin is dark red colored on green yellow base. The fruit flesh is delicate, juicy, firm, crispy and aromatic, and in milk-yellow color. The flavor is good. The soluble solid content is 11.4%, the firmness 9.98kg/cm². The fruits ripen in early October The fruits can be kept to February of the next year under the natural storage condition.

134. 新红星

来源　原产美国，1953年在俄勒冈州发现的红星短枝型全株芽变，为元帅系第三代品种。1965年引入我国。

主要性状　果实圆锥形，果顶五棱凸起明显，果形指数0.9～1.0，端正、高桩；果个较大，平均单果重230g左右，最大果重300g；底色黄绿，全面着浓红色，树冠内外着色均匀一致，鲜艳美观；果面光滑，有光泽，无锈，蜡质较多，果粉薄，果点较稀；风味酸甜，香气浓，微具涩味，可溶性固形物含量13.5%，可滴定酸含量0.25%，果肉硬度7.5kg/cm^2，品质中上；果肉绿白色，肉质较细，松脆，汁多。果实发育期135d左右，较耐贮藏。植株紧凑，适于密植栽培。

134. Starkrimson Delicious

Origin　Starkrimson Delicious, originated from United States, is a spur type mutation variety of Starking Delicious, recognizing as a third generation variety of Delicious strain. It was found in Oregon in 1953 and introduced into China in 1965.

Main Characters　The fruit is conical with angels on the fruit top and an average weight of 230g. Yellow green ground color with full striped strong red over color, good surface finish, glossy, thick waxiness and small dots. The flavor is intensely aromatic and light acerbity, tart-sweet with a soluble solid substance content of 13.5% and a total titratable acid content of 0.25%. The flesh is light green, fine, crispy, very juicy and firm with a firmness of 7.5kg/cm^2. The fruit developing period is about 135 days. Good storability. Suitable for high density planting.

中国苹果品种

135. 新世界

来源　日本群马县农业综合试验场北部分场以富士 × 赤城杂交选育的晚熟品种。1988年获日本农林水产省种苗法新品种登记证。

主要性状　果实长圆形，果个较大，平均单果重200g，果面底色黄绿，着浓红色，具明显条纹；果肉黄白色，肉质致密，脆而硬，果汁中多，风味酸甜，有香味；可溶性固形物含量13% ～ 15%，可滴定酸含量0.3%左右，果肉硬度8.7kg/cm²，品质上等。果实发育期175d，10月上中旬成熟。果实耐贮藏，耐贮性与红富士相同。基本上无生理落果和采前落果现象，早果性和丰产性均较强。抗寒性强于富士，对斑点落叶病、白粉病、轮纹病有一定抗性。

135. Shinsekai

Origin　Shinsekai is a variety selected by northern part of Agricultural Experimental Station in Japan Qunma Prefecture. Its parentage is Fuji × Chicheng. The plant registration certificate of new variety was got from Agricultural Ministry of Japan in 1988.

Main Characters　The fruit shape is oblong globose. The average fruit weight is about 200g. The skin of fruit is striped on dark red base. The flesh is in light yellow color, delicate, firm and crispy. The fruit firmness is 8.7kg/cm². The soluble solid substance content is between 13% and 15% and the total titratable acid is about 0.3%. The fruit developing period is about 175 days, maturing in the mid October. The fruits store very well. There are no physiological fruit drops and re-harvest fruit drops. The trees are precocious and productive. It is resistant to leaf spot, powdery mildew and ring rot diseases. It also has good tolerance to cold stress, even better than Fuji.

136. 旭

来源　加拿大自然实生育成的中晚熟品种，1870年发表。我国于1910年引入。

主要性状　果实扁圆形，平均单果重177.5g；盖色紫红色霞，阳面有不明显条纹。果面光滑，果点小；果肉白色，风味甜酸；可溶性固形物含量11.9%，可滴定酸含量0.72%，果肉硬度7.0kg/cm^2；汁液多，风味浓，适于鲜食。果实发育期110d左右。较丰产。树体适应性广，抗逆性强，较抗寒。

136. McIntosh

Origin　McIntosh is a middle late apple variety selected from natural seedlings in Canada, which was released in 1870 and brought into China in 1910.

Main Characters　The fruit shape is oblate. Its average fruit weight is about 177.5g. The cover color of skin is purple red with unobvious stripes on the sunny side. The fruit surface is smooth with small dots. The flesh is white with sweet and sour flavor. The content of soluble solid is 11.9%, and the content of total titratable acid is 0.72%.The flesh hardness is 7.0kg/cm^2. 'McIntosh' is very productive, juicy with good flavor, and suitable for fresh fruit. The fruit maturation is about 110 days. The cultivar has high adaptability, strong stress resistance especially to cold.

137. 选拔红玉

来源 辽宁省果树科学研究所于2000年从日本引进的红玉早熟酸型芽变。

主要性状 果实近圆形，平均单果重180g；果实着浓红色，底色黄绿；果肉黄白色，风味微酸；可溶性固形物含量16.3%，可滴定酸含量1.14%，果肉硬度8.7kg/cm²；肉质细脆、汁多，味浓、微香，鲜切后褐变轻，果汁风味酸甜浓郁，有香气，为鲜食、加工兼用品种。果实发育期150d左右。选拔红玉抗寒性与红玉相近，对斑点落叶病、苹果树腐烂病和轮纹病等抗性较强。

137. Selected Jonathan

Origin Selected Jonathan is early ripening and sour mutation of Jonathan introduced from Japan by Liaoning Research Institute of Pomology in 2000.

Main Characters The fruit shape is oblate. Its average fruit weight is 180g. The fruit ground color is yellowish-green. The flesh is in yellowish white color. It tastes slight sour. The soluble solid substance content is 16.3%, the total titratable acid is 1.14%, the firmness is 8.7kg/cm². The flesh is fine and crispy, juicy, tasteful and slight aroma. It is a slight browning after fresh-cut. It is juicy strong sweet and sour flavor, aroma. It is suitable for fresh eating and processing. The fruit developing period is about 150 days. Its resistibility of cold is equal to Jonathan. It is resistance to leaf spot disease, apple canker and ring spot.

138. 岩富10

来源　日本从富士芽变中选育的晚熟品种。1979年引入我国。

主要性状　果圆形或近圆形，平均单果重220g；底色黄绿，盖色鲜红；果面光滑，果点中大；果肉黄白，风味酸甜适度；可溶性固形物含量13.7%，可滴定酸含量0.26%，果肉硬度6.9kg/cm²；肉质细密，汁液多，有香味，适于鲜食。果实发育期165d左右。丰产性强。生长势强，适应性不强，抗寒、抗病力弱。

138. Yanfu 10

Origin　Yanfu 10 is a late-maturing apple variety selected from bud mutation of Fuji in Japan, which was brought into China in 1979.

Main Characters　The fruit shape is globose or near globose. Its average fruit weight is about 220g. The ground color of skin is yellow green, and the cover color is turkeyt red. The fruit surface is smooth with medium dots. The flesh is yellow white with moderately sweet and sour flavor. The content of soluble solid is 13.7%, and the content of total titratable acid is 0.26%.The flesh hardness is 6.9kg/cm². Yanfu 10 is very productive with fine, juicy and aromatic fruit, and suitable for fresh fruit. The fruit maturation is about 165 days. The tree has strong vigor, poor adaptability, and poor resistance to cold and diseases.

139. 印度

来源　原产日本，来源不详。20世纪初开始在日本栽培，同一时期传入我国辽南一带。甘甜型果实的代表性品种。

主要性状　果实长圆形，多呈偏斜状；平均单果重188g，最大果重400g；底色绿或淡绿，阳面可带暗红或淡紫红晕；果面较粗糙，蜡质少，果点大、明显，黄褐色或锈色，有果粉；风味甘甜，可溶性固形物含量15.6%，可滴定酸含量0.16%；果肉绿黄色，肉质硬脆，致密、稍粗，汁少，少香气，常有水心病发生；耐贮藏，常温下可存放至翌年3月，贮后肉质松脆，汁液增多，香味增加，品质提高。果实发育期160d左右。连续结果能力较差，有采前落果现象。对病虫害抗性较强。

139. Indo

Origin　Indo, originated from Japan, is an uncertain of the source and a symbol sweetest variety. It was introduced into Japan and Liaonan, China in the early twentieth century.

Main Characters　The fruit is oblong globose with an average weight of 188g. Green or light green ground color with full strong red over color, coarse surface finish, little fruit powder and obvious large dots. The flavor is aromatic, sweetness with a soluble solid substance content of 15.6% and a total titratable acid content of 0.16%. The flesh is greenish yellow, firm, crisp with a little juicy. The fruit developing period is about 160 days. Weak fruiting continuously with fruit drop of pre-harvest. Storage life at room temperature for 5 months. High resistance to diseases.

140. 乙女

来源　辽宁省果树科学研究所1979年从日本引进的鲜食与观赏晚熟苹果新品种，母本红玉，父本不详。2006年通过辽宁省非主要农作物品种备案办公室备案。

主要性状　果实圆形；平均单果重50g；全面着鲜红色，艳丽；果面光滑，有光泽；果肉黄白色，风味酸甜适度；可溶性固形物含量14.8%，可滴定酸含量0.38%，果肉硬度7.7kg/cm^2；肉质松脆，汁液中多，品质中上。丰产性好，串花枝多，枝条较软，适合造型，果实观赏期40d，是一个优良的鲜食兼观赏品种。果实发育期155d左右。树体抗寒性与金冠相当，对苹果轮纹病和斑点落叶病抗性较强。

140. Alps Otome apple

Origin　Alps Otome apple is a new late ripening apple variety with fresh-eating and ornamental. It was first introduced from Japan by Liaoning

Research Institute of Pomology in 1979. It was developed from the cross of Jonathan × unknown cultivar. It was examined and approved by Liaoning Crop Cultivar Registration Committee in 2006.

Main Characters   The fruit shape is round. Its average fruit weight is about 50g. The skin of fruit is in full red color, good surface and glide. The flesh is in yellowish-white color and in good balance of sugar and acids. The soluble solid substance content is 14.8%, the total titratable acid content is 0.38%, the firmness is 7.7kg/cm^2. Alps Otome is very productive, more flower branch, crispy, middle juicy, middle good quality. The branch is soft and made type easily. It lasts 40 days for ornamental. It is a good fresh eating and ornamental variety. The fruit developing period is about 155 days. It is similar to Golden Delicious in tolerant to cold weather. It was strong resistance to ring rot disease and *Alternaria alternate*.

141. 樱桃嘎拉

来源 新西兰从皇家嘎拉中选出的浓红着色类型芽变。

主要性状 果实短圆锥形，平均单果重130g；果实底色绿黄，果面着浓红色条纹；果面光滑，果点大而多；果肉淡黄色，肉质松脆，汁中多；可溶性固形物含量13.8%，酸甜味浓，芳香浓郁，品质上等。果实发育期110～115d，比皇家嘎拉早成熟7～10d。

141. Cherry Gala

Origin Cherry Gala is a dark red mutation of Royal Gala selected by New Zealand.

Main Characters The fruit shape is truncate conical. Its average fruit weight is 130g. The skin of fruit is dark red with stripe and green yellow background color, smooth surface with many and big fruit dots. The flesh is light yellow color, crisp and juicy. The soluble solid content is 13.8%. The eating quality is good with suitable sour sweet flavor and aromatic. The fruit developing period is 110~115 days. The mature period is earlier 7~10 days than Royal Gala.

142. 元帅

来源 源自美国的自然实生品种，19世纪80年代发现于艾奥瓦州，20世纪初传入日本，随后传入我国。是元帅系品种的先祖。

主要性状 果实圆锥形或长圆锥形，果顶五棱凸起；平均单果重200～250g，大者可达500g；底色黄绿，阳面有鲜红色霞和浓紫红色粗条纹，阴面着色较差，果面光滑，蜡质较厚，果点小而密，灰白色，外观美；味浓甜或稍带酸味，有芳香；可溶性固形物含量12%左右，可滴定酸含量0.25%，果肉硬度7.4kg/cm^2，品质上等；果肉淡黄色，肉质松脆，多汁。最适食用期为采后2个月左右，果肉易沙化。

142. Delicious

Origin Delicious, originated from United States, is a natural seedling variety. It was found in Peru, Iowa in 1980s and introduced into Japan and China at the beginning of the twentieth century.

Main Characters The fruit is conical or oblong conical with angels on the fruit top and an average weight of 200~250g. Yellow green ground color with bright red sunglow and rough striped purplish red over color in sunny side and light color in shaded side, good surface finish, glossy, small dots and thick waxiness. The flavor is aromatic, dense sweet with a soluble solid substance content of 12% and a total titratable acid content of 0.25%. The flesh is light yellow, loose, crispy and very juicy with a firmness of 7.4kg/cm^2. The period of optimal consumption is about 2 months postharvest. Easy to sand under the room temperature.

143. 早红

来源　中国农业科学院郑州果树研究所从意大利引进材料中筛选培育而成。初定名为意大利早红。2006年通过河南省林木品种审定，定名为早红。

主要性状　果实近圆锥形，平均单果重173g；底色绿黄，全面或多半面橙红色；果肉淡黄色，肉质细、松脆、汁多；可溶性固形物含量13.0%，果肉硬度7.12kg/cm²，风味酸甜适度、有香味，品质优。果实发育期110d左右，比嘎拉早1周成熟。

143. Zaohong

Origin　Zaohong is selected from materials originated from Italian by Zhengzhou Fruit Research Institute, Chinese Academy of Agricultural Sciences. It was examined and approved by Henan Forest Cultivar Registration Committee in 2006.

Main Characters　The fruit shape is near conical. Its average fruit weight is 173 g. The skin of fruit is full or half of orange red on the green yellow background color. The flesh is light yellow, fine, juicy. The soluble solid content is 13.0%, the firmness is 7.12kg/cm². The eating quality is good with aromatic and good balance of sugar and acids. The fruit developing period is about 110 days. The mature period is earlier one week than Gala.

144. 早捷

来源 美国纽约州农业试验站以Quinte × 七月红杂交选育而成的极早熟品种。中国农业科学院郑州果树研究所1984年引进。

主要性状 果实扁圆或近圆形；平均单果重156g；底色绿黄，着鲜红色霞；果面光滑，果点小；果肉乳白色，肉质松脆，汁液中多；可溶性固形物含量11.8%，可滴定酸含量0.68%；风味甜酸而浓，有香味，品质中上。果实发育期60d左右，为极早熟鲜食品种。

144. Geneva Early

Origin Geneva Early is a very early mature apple variety selected by New York State Agricultural Experiment Station. Its parentage is Quinte × July Red. It was introduced to China by Zhengzhou Fruit Research Institute, Chinese Academy of Agricultural Sciences in 1984.

Main Characters The fruit shape is or near globose. Its average fruit weight is 156g. The skin is bright blush red on the green yellow background color with smooth surface and small fruit dots. The flesh is cream color, crispy and moderate juicy. The soluble solid content is 11.8%, the total titratable acid 0.68%. Its eating quality is medium to good with strong sweet-sour flavor and aromatic. The fruit developing period is about 60 days. It is one of the earliest ripening fresh fruit variety.

束怀瑞，等. 1999. 苹果学[M]. 北京：中国农业出版社.

李育农. 2001. 苹果属植物种质资源研究[M]. 北京：中国农业出版社, 6-9.

李育农. 1989. 世界苹果和苹果属植物基因中心的研究初报[J]. 园艺学报，16(2): 101-107.

陆秋农, 贾定贤. 1999. 中国果树志·苹果卷[M]. 北京：中国林业出版社.

图书在版编目（CIP）数据

中国苹果品种：汉、英／丛佩华主编 . —北京：
中国农业出版社，2015.10
ISBN 978-7-109-20880-3

Ⅰ. ①中… Ⅱ. ①丛… Ⅲ. ①苹果–品种–中国–
汉、英 Ⅳ. ①S661. 1

中国版本图书馆CIP数据核字（2015）第207540号

中国农业出版社出版
（北京市朝阳区麦子店街18号楼）
（邮政编码100125）
责任编辑 张 利 黄 宇

北京中科印刷有限公司印刷 新华书店北京发行所发行
2015年11月第1版 2015年11月北京第1次印刷

开本：889mm × 1194mm 1/16 印张：11
字数：305千字
定价：200.00元
（凡本版图书出现印刷、装订错误，请向出版社发行部调换）